Critical Issues in New Zealand Society

Green Politics

Critical Issues in New Zealand Society

General editor: Paul Spoonley

Green Politics

Stephen Rainbow

Auckland
OXFORD UNIVERSITY PRESS
Melbourne Oxford New York

Oxford University Press, Walton Street, Oxford OX2 6DP
Oxford   New York   Toronto
Delhi   Bombay   Calcutta   Madras   Karachi
Kuala Lumpur   Singapore   Hong Kong   Tokyo
Nairobi   Dar es Salaam   Cape Town
Melbourne   Auckland   Madrid
and associated companies in
Berlin   Ibadan

Oxford is a trade mark of Oxford University Press

National Library of New Zealand
Cataloguing-in-Publication data

Rainbow, Stephen, 1961–
   Green Politics / Stephen Rainbow.
   1 v. (Critical issues in New Zealand society)
   ISBN 0–19–558272–1
   1. Green movement. 2. Green movement—New Zealand.
3. Environmental policy—New Zealand. I. Title. II. Series.
      322.44 (322.440993)   zbn93–032138

Cover designed by Nikolas Andrew
Typeset by Egan-Reid Ltd
Printed in New Zealand by GP Print Ltd
Published by Oxford University Press
1A Matai Road, Greenlane
PO Box 11–149 Auckland, New Zealand

Contents

Editor's Preface

As with earlier books in this series, the intention with this book has been to encourage a social scientist, who has something distinctive to say about an aspect of New Zealand society, to state their argument in a readable fashion. The author of the present book has been asked to do two things. The first is to introduce the reader to the key debates in a particular area—in this case, green politics. The second is to adopt a position and to argue it through. An important part of this requirement is the need to offer a way (or ways) forward. It is not sufficient to simply criticize.

Stephen Rainbow is well positioned to write on green issues. His doctorate focused on such matters and he has personal experience as a key player in New Zealand's evolving green politics, most notably as a councillor on the Wellington City Council. In this book he argues that the anti-modernist arguments of contemporary greens are the direct descendants of a variety of nineteenth-century romantic and utopian movements. In a New Zealand context, these arguments are critical because they reinvest an importance in non-material aspects of social life and offer a critique of current models of development and growth which rely upon a narrow conception of economic factors. His response and his suggestion for a way forward are essentially pragmatist in that he argues for the State to take a responsibility for a sustainable future, and for green activists to adopt achievable and credible strategies for change. In offering such arguments, the author has unequivocally met the basic requirements of the series.

Paul Spoonley
Series Editor

Introduction

Modern society appears to be confronted with a host of insoluble problems. At every level—political, social, economic, and environmental—existing institutions appear unable to grapple with the complex issues raised by persistent change. Momentous events, such as the collapse of the Soviet Union on the world stage, are paralleled on a personal level by the trauma experienced by many individuals as a result of redundancy or unemployment.

The environmental crisis, the result of human violence against natural surroundings, is paralleled by violence between humans, whether within the family or between feuding nationalist groups.

Inevitably, this degree of change over a number of years has been reflected politically. The main result has been a widespread disillusionment with traditional political institutions, political processes, and politicians. There has also been an upsurge of new political movements. Not all of these new movements are positive, as the growth in extremist anti-immigrant movements in Europe and the rise of religious fundamentalism demonstrate. Some movements respond to the uncertainty of the present by seeking a return to the certainties of the past. Others aspire to shift the political agenda so that it may adapt to the challenges of the age in a more coherent and future-oriented way. Green politics belongs to this latter category, and this book is devoted to the task of analysing green politics. Such analysis is valuable not only because of the nature of the green response to the multifarious crises of modern society itself, but also because green politics offers potentially valuable pointers to the future direction of modern societies.

Green politics has a unique appeal which has gained the support of a significant minority of people in most Western democracies. It has been central to the de-alignment of existing political allegiances and structures in modern industrial nations. While its political success has been erratic in some countries, green politics has been described as the most important political innovation since the advent of socialism (Porritt and Winner, 1988). Green politics claims to represent a fundamental shift, or 'turning point', in the direction of modern industrial societies, hence its description as a 'politics of hope' (Blackwell and Seabrook, 1988). Green politics is built around the notion that the plethora of modern 'crises' are interrelated and stem from the same sources. These root causes must be addressed if solutions to the crisis of modernity are to be found.

Green parties have been the most common vehicle (although not the only one) for articulating green politics. While frequently resisting the trappings and appearance of what might conventionally be expected of a political party, green parties are an established element of the electoral landscape in most modern industrial societies. Green parties are (or have been until recently in the cases of Sweden and Germany) represented in the parliaments of most democratic nations. Green politics has also been a key component of various change-oriented movements in the former Soviet Union and the Eastern bloc, where environmental issues became a major focus of activities in opposing the former Soviet regime. In some nations, green parties have held the balance of power, as in Tasmania, and in several German state-level and city governments. Even where they have gained little or no representation in established institutions, green parties have influenced political agendas. For example, the world's first national green party, New Zealand's Values Party, was articulating nuclear-free and gender-equality issues in New Zealand long before the established parties embraced such stands.

The influence of the agenda of green politics is demonstrated by the efforts of established political parties and their leaders to adopt the green mantle and to respond to obvious growing public concern about green issues. Green issues—whether fighting nuclear power or opposing the extension of motorways and airports—have become potent political issues. A panoply of organizations has formed to mobilize people around green issues, whether specific local issues or broader national and global agendas, to ensure a viable planetary future. These organizations have provided an important constituency for green politics, and can be differentiated from green pressure-group activity by their focus on seeking political power with a broad range of appropriate policies. Green politics gained a significant following during the 1980s and green issues promise to be one of the major sources of political division as the 1990s progress.

A part of the ability of green politics to influence the political system is its vote-pulling power. Green parties have gained credible percentages of the vote even where electoral systems have blocked minor party representation in the national parliament, as in New Zealand. Green parties have successfully challenged established party systems even in stable political environments like that of Sweden. On their election to the Riksdag in 1988, the Swedish Greens became the first new party to enter the Swedish Parliament for seventy years. It is the German Greens, however, who are best known internationally. Amid a blaze of publicity, they entered the Bundestag in 1983 with twenty-eight seats. They increased their seats to forty-two at the 1987 elections and, while they lost their parliamentary representation in 1990, they still hold the balance of power at a local level in several German cities and states.

Green parties are characterized by their articulation of new themes, their rearticulation of old, forgotten themes, and the extent to which they reinvigorate the programmes of established parties (Papadakis, 1989). Green parties are programmatic; their primary role, Papadakis (1989: 64) asserts, is as a 'policy making agency'. Green policies represent what has been described as a 'new politics' agenda, comprising issues which are ignored or marginalized by conventional politics. A 'new politics' agenda may be present in other parties, most often on the left of socialist or social democratic parties. Established parties, however, often attempt to incorporate new issues, like the environment, into the established Left–Right political spectrum. The Swedish Social Democratic Party, for example, responded to growing environmental concern by starting a campaign which focused on the '*working* environment'. In contrast, greens see their agenda as crucial in its own right, and often claim that green issues are simply tacked on to the existing programmes of established parties.

While mainstream parties have now adopted many green policies, this does not mean a 'similar approach to politics nor shared aims, values and direction' (Papadakis, 1989: 70). Green parties are often distinctive because of the degree of moral intensity which they attach to particular issues. Green campaigning tends to be generalized and moralistic in nature, and strongly connected to demands for increased participation in political decision-making. Questions of political process have a prominence in green politics which is rarely seen in other political ideologies, although there are close parallels with the practices in many of the new social movements, e.g. the women's movement.

This book will attempt to explain those factors which make green politics unique. It will also examine the record and prospects of green politics. Having borne the first green party in the world, New Zealanders are in a unique situation to examine two decades of green politics. Yet contemporary green politics in New Zealand, as elsewhere, appears not to have learnt from history. After more than twenty years of green politics, it is still difficult to identify a clear green vision of the preferred future and the means for its achievement.

Green Politics in New Zealand

New Zealand has a unique place in the history of green politics. The world's first national green party, the Values Party, was formed in Wellington in May 1972. The fledgling Values Party confounded the pundits by winning 2.8 per cent of the national vote at the 1972 elections. Values gave a new emphasis to issues such as energy, the environment, and the equality of women. These issues directly reflected many of the concerns of the new social and political movements which had arisen in New Zealand (and throughout the modern world) during the 1960s. The 1960s witnessed the breakdown of the postwar consensus in New Zealand.

For the first time, there were major divisions among the public over a foreign policy issue. New Zealand's involvement in the Vietnam War caused a great deal of discord, particularly among young people. As well as anti-Vietnam War protests, the 1960s also witnessed the first national environmental issue in New Zealand's history. This concerned the plan to raise Lake Manapouri for hydroelectric development. Opposition to the raising of Lake Manapouri struck at the heart of the development ethic which had driven New Zealand's colonial history. Hydroelectric development was crucial to the State's vision of New Zealand. Cheap hydroelectricity was seen as a natural advantage on which New Zealand could capitalize to overcome the disadvantages associated with being so distant from our major markets. Cheap electricity would provide the energy for a programme of industrialization which would finance the welfare state and an improved standard of living (Rainbow, 1992). The Ministry of Works' dam-building projects in the postwar years were a major source of economic activity, providing many jobs as well as the energy for an emerging industrial sector.

By the 1960s, there was no longer the previous consensus about the acceptability of the costs of industrialization. Increased levels of education, growing affluence, and the advent of a critical media, including television, were among the reasons that the postwar consensus in New Zealand collapsed (Chapman, 1981). Significant numbers of people in New Zealand also shared a growing concern with lifestyle issues. Personalities, like the poet James K. Baxter, touched a chord with many young people. They shared his contempt of suburban life and the forty-hour working week as the central characteristics of modern society. Alternative lifestyles became popular, even gaining official support when the Third Labour Government introduced a scheme of giving marginal State land over to communities which wished to practise kibbutz-type lifestyles.

For a time the Values Party was able to give political expression to these alternative ideas and views. But Values was denied parliamentary representation by a first-past-the-post electoral system which ensured the dominance of the two major parties. Lack of electoral success was one of the reasons that Values suffered from debilitating internal conflicts by the late 1970s, as it decided on its future role in the face of the permanent obstacle created by the electoral system. Many party activists subsequently chose to put their energies into single-issue groups such as the movements to ensure New Zealand did not adopt nuclear power (the Campaign for Non-Nuclear Futures) or into the 'Repeal' petition for liberalized abortion laws. During the 1980s, Values stood a small number of candidates in general elections and continued a minimal party structure, but its effectiveness as a national political force had ended along with the 1970s.

In the run-up to the 1989 local elections, groups from around New Zealand, including the remnants of the Values Party, met in Wellington to discuss the formation of a new green party similar to those which were by this time gaining much attention in other parts of the world. Green candidates subsequently stood in local elections for the first time in 1989, with some successes. Over the following year, the Green Party of Aotearoa/New Zealand was formed in time to contest the 1990 general election. The Greens won 6.9 per cent of the total vote, but no parliamentary representation. Frustrated by the inability of the fledgling party to capitalize on its percentage of the vote, the party's hierarchy subsequently moved to incorporate the Greens into a coalition of third parties.

This decision was endorsed by the Green conference in May 1992. What began as a minor-party coalition to promote electoral reform prior to the September 1992 referendum on electoral reform, ended up as a fully-fledged third party called the Alliance. It is therefore unlikely that voters will have a Green option on the ballot paper for the 1993 elections. The assumption that green voters will transfer their allegiance to an alliance of third parties sharing little more than their minority status is, however, a dubious one. The future of a uniquely green presence in New Zealand politics therefore appears uncertain.

Assumptions Behind This Book

It is useful if some of the assumptions underlying this book are made explicit at the outset. These assumptions arise out of an examination of the history of green politics both in New Zealand and abroad, and from my personal involvement in green politics. As well as working with green parties in Finland and Sweden during the preparation of my doctoral thesis (on green politics), I have served as a Green city councillor in Wellington since 1989. Some assumptions upon which the arguments presented in this book are based include the following: Those in green politics must

- be concerned with more than the environment
- think through the practical implications of green policies
- think more strategically
- acknowledge that politics will be increasingly eclectic in postmodern society
- place at the forefront of their agenda the urgent need for a democratic consensus on solutions to the environmental crisis
- learn from modern political history in their future political designs.

Scope of Green Politics

At the core of green politics is a general critique and rejection of the direction of modern industrial society. The environmental crisis is simply the most conspicuous example of the negative effects of the values which underpin modern industrial society. These values are inappropriate as humankind takes an unprecedented role in determining the evolution of the entire planet. By encouraging each decision to be made with an awareness of human responsibility for global evolution, green politics is steeped in a moralism which is unusual in contemporary politics and which has an impact on every aspect of public policy.

A rejection of industrialism, as much as a concern for the environment, lies at the core of the green ideology. Industrialism, in both its communist and capitalist forms, has been central to the direction of the modern society with which greens feel such discomfort. But industrialism refers to more than a particular mode of production: it represents a total culture, an entire paradigm. The rejection of industrialism leads to a variety of responses within the contemporary green movement. Splits within the greens gravitate around fundamentally opposed pre-industrial or postindustrial models of the future. These divisions reveal that a concern with the environment does not, of itself, dictate a particular political position or programme.

Practical Implications of Green Policies

There is a need to examine the practical implications of implementing green policies. While attempts to analyse the green ideology have been made, very little has been written about how a green society might look. Even less has been written about the tactics and strategies necessary for the attainment of a greener society. This book attempts to address this vacuum. It will suggest practical ways in which the possibility of a concerted green contribution to a postmodern society might be pursued. It will raise issues which have been inadequately canvassed, such as the cultural and political limits to the achievement of the green agenda.

Specifically, there is a need to discuss the extent to which the central green tenet of sustainable development is compatible with growth-oriented, privately controlled industrial economies. Other issues which need to be addressed include whether or not democracy is expendable in pursuit of the protection of the environment, and whether or not the moralism of green politics is compatible with a pluralist postmodern society. Furthermore, greens need to openly discuss whether a green future will require a new formulation of the human psyche akin to communist attempts to create the *New Soviet Man* after 1917. These and other questions need answers as people require a clearer picture of exactly what green politics means *in practice*.

Strategic Thinking by Greens

Green politics has the potential to contribute to an eclectic postmodern political future in modern industrial societies. But green politics also needs to recognize the limits to political action within liberal democracies and to differentiate between what is achievable at the level of civil society and at the level of the State. This involves the need for greater strategic thinking by greens, both about what their primary goals are, and how they might be achieved. For example, how is sustainable economic development to be achieved? Will green policies necessarily mean an extension of public powers? How can a generalized unease with modern society be turned into the realization of a political consensus on the need for decisive action at local and global levels in order to address the global environmental crisis?

Fundamental to these strategic issues is the green view of power. Green approaches to power typically call for its redistribution to an active citizenry at the grass roots. Calls for the decentralization of decision-making along with a general remoralization of politics and a concern for the environment do not of themselves dictate a political programme for the future. Typically, green movements include a spectrum of opinions ranging from anti-democratic fundamentalists to ultra-pluralists from the new social movements. There is a need to devote greater attention to how change occurs and how political power might be used to achieve green goals. A consensus simply that conventional action is undesirable is insufficient as a basis for a political programme (Pakulski, 1990: 171).

Green politics has a much clearer idea of what it is opposed to than of what it is for. Yet to motivate people concrete alternatives and 'pieces of a workable future' must be presented to a public far wider than the minority holding the 'new values' upon which green politics is predicated. Many of the necessary solutions to global environmental problems already exist. What is required is a policy-making framework involving both incentives and legislation to ensure that such solutions are implemented as part of a concerted direction of political energies towards the attainment of sustainable development. New Zealand is in an ideal situation—because of its relatively unspoilt environment, its size, and physical isolation—to introduce precisely this kind of policy regime, and green politics should be at the forefront of efforts to ensure that this happens.

Another of the main roles of the greens should be to publicize and promote working examples of how things can be done differently, in a socially and environmentally sustainable manner. I reject the 'grand narrative' of a green fundamentalist utopia and the catastrophic predictions of the doomsdayers, preferring the more optimistic appraisal of those who hope that concerted human effort and responsive political structures can help to reverse current destructive trends.

Politics in Postmodern Society

Any political writing includes assumptions about the past, present, and future. This book is based on the assumption that the postmodernist description of a future society marked by increasing pluralism is both desirable and likely. Consequently, future political designs, and personal values, behaviours, and spiritual beliefs, are likely to be increasingly diverse. A pluralist postmodern future does not preclude the need for a consensus on addressing the environmental crisis, however, and may, in fact, depend upon it. Without such a consensus, there is the potential for environmental problems to become a source of conflict within nations as well as between them.

The assumption here is of progress towards a new stage of social evolution, from the mass solutions of the modern industrial age to a pluralistic *post*modern era based on an acceptance of the equal validity of a multiplicity of individual tastes and desires. This will be a society marked by an acceptance of a wider range of norms and values than has been evident during the industrial era. Politically this is manifested by the challenges to universalistic schemes and utopias such as Soviet communism. Similarly, green politics can be interpreted as a challenge to the pervasive ideology of acquisitive materialism in Western countries.

The green movement will be one of many different movements seeking to influence policy agendas and the way people live in a postmodern society. In such a pluralist environment, mutual respect and individual rights will be of paramount importance. It is, therefore, vitally important that green politics recognizes the likely shape of future political designs and abandons any universalistic ambitions it harbours. Greens must be committed to working as effectively as possible within a democratic, pluralist framework which respects the rights of individuals as well as those of the planet.

Democratic Consensus on Environmental Solutions

While green politics is concerned with more than simply the environment, there is a need to focus on achievable objectives. A primary role for green politics must, therefore, be to work towards achieving a democratic consensus around the need to take action to alleviate the global environmental crisis. Unless current patterns of destruction are reversed, assumptions about increasing pluralism and the possibility of a broader distribution of an increased quality of life could well become irrelevant. Without concerted international cooperation in addressing the environmental crisis, there is a real possibility of resource and environmental issues being used as a justification for diminishing democratic processes in the future.

The reform of political institutions so that they 'learn' from the negative feedback signals of social malfunctioning and environmental degradation will be crucial to the global future. Green politics must be at the forefront of attempts at institutional reform, ensuring that currently dominant economic values are countered by a system of governance which takes a holistic approach to decision-making and which includes an awareness of the true costs of development. A holistic approach will acknowledge the importance of 'internalizing' factors such as the environment and the quality of life which are too often excluded from decisions made in contemporary political systems.

Future Designs and Modern Political History

This book suggests practical ways in which green politics might realistically work towards achieving its goals. This approach is based, in part, on the assumption that the price of separating theory from action has been too high during this century. This has led, for example, to support for Soviet bloc regimes from 'progressive' Westerners even while the most heinous crimes were being committed in the name of socialism. Any future political designs simply cannot ignore the lessons of modern history, which include appalling suffering and genocide in the name of some higher ideal, be it National Socialism or Communism. Green anti-hierarchicalism and decentralism is one response at the level of political organization, but the required response at a broader ideological level, which must include the rejection of utopian schemes for universal application, has not been followed through. There are still elements within green politics which see democracy as dispensable in the pursuit of environmental salvation. It is quite conceivable that 'saving the environment' could be just the kind of potent ideal which could be used to legitimize governance of a less than benevolent kind in future.

Format of the Book

The first chapter locates green politics in a historical context, arguing that green politics represents a continuity of anti-modernist ideologies which have been in existence throughout the modern era. Chapter Two explains what is unique about the contemporary green ideology, and Chapter Three examines 'sustainable development', a concept which is at the heart of the green ideology. Sustainable development, it is argued, poses a fundamental challenge to existing forms of political and economic organization. This makes the need for green strategies and policies, the subject of Chapter Four, all the greater. This chapter looks at the kind of changes which might occur at all levels of society if the green contribution to a postmodern and increasingly pluralistic society is to be effective. The fifth chapter looks at the likely future shape of modern industrial society, arguing that a postmodern epoch is emerging which poses unique opportunities—and challenges—for green politics.

1 The Historical Context

As the promotion of rationality to the exclusion of alternative criteria of action, and in particular the tendency to subordinate the use of violence to rational calculus, has been long ago acknowledged as a constitutive feature of modern civilisation—the Holocaust-style phenomena must be recognised as legitimate outcomes of the civilising tendency, and its constant potential. (Bauman, 1989: 28)

This chapter places the contemporary manifestations of green politics in a historical context. As well as arguing that green politics represents the continuity of an anti-modernist tradition, it will also explore some of the issues which have led to the recent upsurge of green politics. In particular, the connection between green politics and the 'New Left', the new social movements, and the early utopian and romantic socialists will be assessed. Even a cursory glance at these movements reveals striking parallels with the concerns articulated by contemporary green politics. This is an important observation because it refutes the claims of those greens who argue that their movement is an ahistorical one without political precedent. In fact, green politics represents the continuity of an anti-modernist tradition which has arisen in a variety of forms in response to 'modern society' (defined by Giddens (1990: 1) as the unique modes of social organization which emerged in Europe from the seventeenth century onwards).

The reasons for attempting to locate green politics on a continuum of anti-modern movements are twofold. Firstly, locating green politics on a continuum of anti-modernist activities legitimizes the claim that green politics is concerned with all facets of modern society, not just the environment. While many individual activists and voters might come to contemporary green politics through environmental issues, the environment cannot be dealt with in isolation. Environmental outcomes are the product of the whole gamut of economic and social policies, as well as of values and cultural patterns. Nor can 'the environment' dictate a particular political programme. Both eco-fascism and free-market environmentalism claim to be equally valid responses to 'environmentalism', even though they are poles apart politically.

My desire is to ensure that green politics is not used as a vehicle for socially regressive policies or for a diminution of freedom or democracy. Establishing green politics' rootedness in a tradition of anti-modernist movements committed to social emancipation and the extension of democracy—as well as to environmental concerns—assists with this task. From a purely environmental perspective, for example, the city can be seen simply as an ecological 'drain' down which endless resources are poured. But this is only a partial perspective on the city, and one which leads to a banal anti-urbanism. It does not constitute a sound basis for addressing the role which cities must play in a sustainable future. Historical, cultural, and social perspectives are also required for a balanced assessment of urbanization, or of any other 'issue'. This point becomes especially important when we explore examples of where environmental arguments, devoid of a social or political context, have been used for untenable ends. For example, the current debate on homosexual law reform in Tasmania has seen 'environmental' metaphors used to describe homosexuality as a pollutant which should be eradicated like toxins from the environment (Croome, 1990). An awareness of green politics as part of a broader tradition of movements concerned with modernity in all its manifestations (including its attitudes to sexuality, for example) can avoid environmental concerns from being abused and misused.

Secondly, showing the historical antecedents to green politics illuminates the difficulties which other anti-modernist movements have faced in the past, difficulties from which contemporary green politics might learn. Significant social and political change is not easily achieved. The often articulated idea that green politics represents a movement and an ideology 'whose time has come' leads to a simplistic belief in the inevitability of green change which the experience of similar movements in the past should quickly dispel. The rejection of the negative costs of 'progress', felt so acutely by a green minority, does not represent the aspirations of the majority of people in modern industrial societies. Green strategies must take cognizance of this fact.

The Historical Background

In New Zealand, as throughout modern, industrial societies, contemporary green politics is rooted in a wide range of movements and issues which have arisen in response to what Galtung (1988: 150) has described as the 'malfunctioning of the Western social formation'. The post-World War Two sources of green politics have been closely related to changing socio-economic factors. Increased affluence in the 1950s and 1960s, and the change in the workforce away from manual and land-based jobs towards the urbanized tertiary sectors, led to changed lifestyles and increased educational opportunities (Baker, Dalton, and Hildebrandt, 1981). These developments amounted to a change in the class basis of society and, in particular, to the growing prominence of the 'new middle class' of skilled workers employed outside the conventional industrial sectors. Frequently, this 'new class' is employed in the public service in such professions as teaching, social work, and health (Eckersley, 1989). In addition, increased levels of education and a greater dissemination of knowledge through the mass media have led to a more informed public which, in turn, demands a greater participation in the decision-making process of their society. Demands such as increased opportunities for political participation, and a defence of quality of life concerns, have consequently been central to the 'new politics' agenda of new social movements, including green parties.

Postmodernist writers Heller and Feher (1988) identify three distinct generational phases since World War Two: the existentialist, alienation, and postmodernist generations. Each phase, they argue, continues 'the pluralisation of the cultural universe in modernity as well as the destruction of class cultures' (Heller and Feher, 1988: 136). These factors have contributed to changing political views and expectations and, in particular, to the creation of a 'new politics' agenda. The 'new politics' agenda emphasizes a participatory model of democracy as well as the importance of personal growth and a rejection of lifestyles based solely on the pursuit of materialism. These concerns depart from the 'old politics' agenda with its emphasis on personal and strategic security promoted through collective organizations ordered around class interests. 'New politics' issues tend to be organized more anarchically and are based around cleavages other than class, including ethnicity and gender, or around specific issues such as peace and environmental issues. The vast array of movements formed around various aspects of the 'new politics' agenda has led to the creation of numerous centres of power outside of traditional parliamentary politics.

Established parliamentary-oriented political parties found it difficult to respond to the new issues and aspirations which emerged during the 1960s and 1970s. Parties of the Left, like New Zealand's Labour Party, which were traditionally looked to by reform-oriented voters, were seen either to have lost their reforming zeal or to have betrayed their fundamental principles altogether on gaining power. Social democratic parties were committed to State-oriented societies and to models of economic development which, while successful in generating increased wealth during the post-World War Two years, were also the source of many of the issues to which environmentalists and others were responding. In addition, the bureaucracies which Leftist parties had nurtured and expanded were perceived as unresponsive to innovation and as obstacles to spontaneous reform.

The despoiling of nature in the name of 'progress' became increasingly unacceptable, as examples like the battle to save Lake Manapouri in the late 1960s and early 1970s show. Energy projects in particular, whether nuclear power in Europe or hydroelectric 'development' in Tasmania and New Zealand, became the focus of mass opposition movements. Typically, the concerns of these movements were not represented by established political parties. The Save Manapouri Campaign showed, for example, that 'the existing political parties seemed to have been an inadequate channel for the articulation of the feelings and sentiments of large masses of people in an area of public affairs which is of great importance' (Cleveland, 1972: 35). This breakdown in the relationship between the established parties and significant numbers of voters has been described as 'linkage failure' by Lawson (1988), and has been central to the rise of green parties.

The 'new politics', a reduction in class voting, and the loosening of traditional partisan ties resulted in a realignment of the electorate based on the new middle class and the young (Dalton *et al.*, 1984). This places those groups among whom greens gain their greatest support at the centre of contemporary political developments. These are the groups most likely to challenge traditional ideas about the welfare state, to question the benefits of continuous technological innovation, to embrace new perspectives on foreign policy, to demand increased participatory rights, and to attach significant value to environmental and quality of life concerns. These issues are precisely those typically embraced by the green political agenda.

But these demands have been expressed in the political arena before green politics, particularly by the New Left during the 1960s. The similarities between the concerns of the New Left and contemporary green politics are hard to discount. This is true not just of policy issues, but of a whole approach to political processes and to political power. Both movements emphasize political means which correspond to their desired ends, and consciously shape their internal political organizations to reflect their commitment to the principles of non-hierarchicalism and participatory democracy generally. In order to demonstrate the extent to which contemporary green politics reflects the aspirations and experience of other movements from recent history, it is worth making a brief examination of the New Left. For, like the New Left, and in spite of its claim to be 'neither left nor right, but in front', green politics addresses many of the issues which the utopian and romantic elements of the socialist political tradition promoted before they were subsumed by the dominant Left strands of 'workerism' or 'labourism'.

The New Left

By the 1960s and 1970s, established parties of the Left had became identified with the establishment in industrial societies and had abandoned any vision of an alternative political or social order. Parties like the New Zealand Labour Party had largely lost any role they might previously have held at the cutting edge of progressive political change. Labour's stance from the 1950s until the 1980s was based largely on the defence of its past creations, namely the welfare state. This defensive stance reduced the ability of the Labour Party to address the issues raised by the 'new politics' agenda. Labour failed to inspire those concerned with issues such as the quality of life, the environment, and the domination of all aspects of life by monetary values. In an Australian and New Zealand context, the respective Labour Parties had become barely distinguishable from their traditional conservative opponents.

The New Left sought to rejuvenate the moribund Left by reviving the idealist, humanist, democratic, and culture-critical elements of socialism. Its policies focused on self-management and the decentralization of power in contrast to the bureaucratic model of the established Left. The New Left rejected the 'stupefying routines' (Katsiaficas, 1987: 5) of much modern life and sought to redefine politics in its classical sense as embracing a concern with the totality of existence. The New Left directly challenged the established Left with an agenda based on personal empowerment and the maximization of spontaneity and experience.

Whereas the traditional Left had emphasized issues of control and ownership of the means of production, the New Left directed its critical attention towards the ideas and values of the culture which underlay modern, industrial society. It addressed the need to restructure daily life by building a counter-culture which sought non-alienated work and leisure, a sense of community, extra-parliamentary political activity, and the human control of technology. Together, these projects comprised 'an alternative society strategy of change' (Young, 1977: 59). Building a better society for future generations, in the established Left mould, gave way to an emphasis on living in the present, and '. . . incarnating the new social relations in the here and now' (Howard, 1989: 103). This strategy reflected the awareness that power was exercised not just by the State and large institutions but also in personal relationships on a day-to-day basis.

Participatory democracy and the New Left's libertarian distrust of State power, parties, leadership, bureaucracies, and representative government took the place of the established Left's commitment to State paternalism and the adversarial pursuit of power (Caute, 1988: 20). A 'natural State' free from restrictions was sought to create 'a space where the free play of the imagination and the work of the hands and mind can find new unity' (Katsiaficas, 1989: 100). Change would come from below, where a politically active population would withdraw its legitimizing power from the hierarchies of domination and the system of false needs and illusory freedoms they represented.

Psychology provided important intellectual foundations for the New Left's cultural critique. Social change and personal liberation merged into one as Marxist theories of alienation were fused with the problem of individual anomie in modern society. The need for a new human character willing to reject authority and convention and to develop personal awareness and autonomous conduct was proclaimed. Fromm's (1976) message was that it was the world that was insane, not the individual. Sanity therefore lay in rebellion and change. Marcuse and others confirmed that it was oppressive social structures which were responsible for the destructive aspects of human nature rather than any intrinsic human characteristics (Young, 1977: 40).

These concerns led to a focus on the detail of everyday existence, because 'the question of morality was bound up with lifestyle, and living in a new way' (Young, 1977: 40). This resulted in a proliferation of attempts at alternative ways of living, including the creation of ohu (rural communes on government land) in New Zealand. The concern with the structure of daily life was also given new emphasis by the emerging women's movement, which drew substantial crowds to its first national conferences in the early 1970s.

The alternative to everyday life as an 'automaton' or 'one-dimensional man' was to demand the right 'of self-management in all spheres of life, self-activity and self-expression' (Howard, 1989: 113). In an attempt to prefigure the desired society, the counter-culture practised non-hierarchical power structures and sought to recognize that people relate to society through a variety of roles, not just through their work in the world outside the home. Civil society, rather than existing power structures, became the focus of alternative models of power and daily living. Groups ranging from the Organization to Halt Military Service (OHMS) to the Campaign Against Foreign Control in New Zealand (CAFCINZ) gave expression to the new-found belief in 'the power of the people' and the connections between individual behaviour and collective outcomes.

Howard (1989: 107) writes that: 'The intellectual roots of the New Left were largely a rejection of the traditional, rationalist views; they were part of a political practice, a social regrouping, and a personal will and moralism'. But the New Left was also inspired by concrete events. The Vietnam War was the most conspicuous of these events. It raised doubts in the minds of many of the younger generation about the morality of their society. It also coincided with the first concerted burst of environmentalism and with indigenous and minority rights movements such as the American Civil Rights Movement and the Gay Liberation Front in Auckland.

The 1960s saw environmental issues gain significant publicity for the first time. Rachel Carson's *Silent Spring* raised widespread concerns about the effects of toxins in the food chain after its publication in 1962. By the end of the 1960s, the 'Limits to Growth' body of literature predicted resource depletion and environmental catastrophe on an unprecedented scale. At local and national levels, rising concern about the environment was apparent in the media. Energy issues were frequently to the fore, whether over the use of nuclear power in America and Europe or the continued destruction wrought by hydroelectric development in Tasmania and New Zealand.

Campaigns for the rights of indigenous peoples found a newly sympathetic response from among the counter-culture. Romantic notions about the statelessness and spontaneity attributed to the lifestyles of indigenous peoples motivated parts of the counter-culture. Indigenous people were seen to possess a symbiosis with the natural environment which had been lost by modern, urbanized people. A return to Mother Earth, de-urbanization, and a world where the joy of life would be unleashed and celebrated were the hallmarks of this late twentieth-century romanticism, expressed through the New Left.

The New Social Movements

As the New Left and the counter-culture became more fragmented and depoliticized after the heady days of the 1960s, people began to direct their energies into more professionalized and specialized groupings. The resulting 'new social movements' tended to be more focused than those of the counter-culture. They included, for example, the womens', peace, and environmental movements. The New Left and the counter-culture had failed to achieve the revolutionary potential of their alternative ways of living envisaged by one of their prophets, Herbert Marcuse (Caute, 1988: 30). But new prophets, such as Touraine and Habermas, saw the new social movements as the prime repositories of social change, replacing existing class-based collectivities with non-class concerns, new organizational methods, and the rejection of programmatic goals. Reflecting their continuity with the New Left—and prefiguring the shape of emerging green parties—the new social movements were utopian, spontaneous, had amorphous memberships within non-hierarchical structures, and sought transformation at a cultural level as well as at the level of the State.

Like the New Left, these new social movements challenged the established Left's notion that economic conditions determine ideas and attitudes. The new movements argued, in contrast, that modes of production arise out of the beliefs, ideas, attitudes, and values which dominate in a given culture. Hence, the need to begin the change process at the cultural level, including at the level of the individual, was crucial.

Compared to the old collectivities such as trade unions and class-based political parties primarily involved in defending their members' interests, the new social movements are future-oriented and transformative. Rather than focusing on a specific adversary (for example, capitalists), the new social movement project, in its totality, amounts to a 'refusal of the social order' (Jennett and Stewart, 1989: 6) or, in other words, to a generalized rejection of modern society. Through various specific acts of refusal, withdrawal, and creating a counter-culture, new social movements are involved in the rejection of old ideological forms of struggle, while collectively representing the potential for a qualitatively different kind of future.

Green politics unites, beneath its umbrella, a variety of new social movements disgruntled with modern society and what has been described as the 'poverty of progress' (Miles and Irvine, 1982). These groups tend to focus on the costs or adverse effects of affluence and the modern infatuation with material progress, including the breakdown of traditional social arrangements and the destruction of the natural environment and cultural heritage. Green politics provides a political voice for these movements, which are extra-parliamentary by nature. Often such social movements direct their activities in a variety of directions, attaching no particular status to established political institutions such as the State. Protesting against nuclear power in a European context, or against illiberal abortion laws in New Zealand, generated significant activism which was frequently not represented by established political parties or institutions. Specific campaigns often foreshadowed the rise of green politics, as these new social movements became frustrated with the unresponsive nature of existing political institutions.

While such movements are crucial to the rise of green politics, the relationship between them and green *parties* has not always been easy. There are often conflicts between these movements which seek immediate and limited goals, and the longer-term and more generalized changes sought by green parties. Another problem in the New Zealand context, exacerbated by the local electoral system, has been the lack of political support which new social movements are able to give green parties because of their need to maintain favour with the established major parties. For example, while the Campaign for Non-Nuclear Futures and 'Repeal' (abortion law reform) petitions were largely circulated through Values Party networks, these single-issue movements were unable to openly support the Values Party because of the need for them to be seen to by the established parties as 'non-partisan'. Nevertheless, the thousands of new social movement networks which exist in most modern nations are crucial to providing an organizational base and a constituency for green politics.

In New Zealand, the connection between the New Left, the new social movements and green politics is clear. The Values Party, formed in 1972, was a direct response to the arid New Zealand political scene and, in particular, to the inability of the Labour Party to adapt to meet the demands of a growing constituency of young, well-educated people whose concerns were distinct from those of their parents. Tony Brunt (1973), founder of the Values Party, claimed that he was one of many young people lying under the tap marked Labour, waiting for a drop of moral leadership (which never eventuated). It was, therefore, not surprising that a new political movement would arise to address the 'new politics' agenda largely ignored by the established parties.

Romanticism and Utopian Socialism

So far I have argued that green politics has grown out of the politically tumultuous period since the 1960s. But green politics is but the most recent of the anti-modernist and anti-industrial movements which have arisen throughout the modern era. Gould (1988) argues that the most fecund period of green politics prior to the present was during the 1890s. This was the time when the 'Back to Nature' movement gained considerable momentum in Britain in response to the rapid process of industrialization.

Green politics is also manifested in other attempts to create a qualitatively different kind of society by those out of step with the dominant spirit of the age. Modern green politics is an extension of the ethical or utopian strands of the early socialist tradition. Galtung (1988: 159) asserts that 'many of the tasks taken on by the Green Movement can be seen as parts of the socialist programme, the preceding wave of social energy left unsolved'. Modern greens attach great importance to environmental issues, but the utopian socialists were also clearly moved to defend nature against 'the dark satanic mills'. Both share a repugnance at the subjugation of all aspects of life to commercial values, as well as to the 'commodification' of even the most intimate aspects of personal life.

The utopian socialists were a significant part of the early socialist tradition. They arose out of an unease with industrialization and adopted a cultural critique of industrialization which emphasized its human and environmental costs. Figures like William Morris, during the latter part of the nineteenth century, opposed the materialism upon which industrialism was predicated, and decried the separation of work from art. They bemoaned the dulling of the emotional and spiritual aspects of life which went hand-in-hand with the spread of the factory system. Mass production and regulated working lives did little to encourage spirituality or creativity. William Morris expressed dismay at the fact that industrialism meant that people had chosen ugliness over beauty and were compelled to work joylessly in the production of useless goods (Morton, 1979: 39).

Industrialization was built upon science and rationality; the emotional, the intuitive, the sensual were devalued. People, like the world around them, were perceived in mechanistic terms because of the new emphasis given to science. The scientific paradigm also helped to legitimize the factory system. Its many claims included the ability to predict, without doubt, personal and planetary behaviour on the basis of 'scientific laws'. The mystical and enchanted aspects of life were invalidated as primitive and subjective. Like the industrialist Gradgrind in Dickens's *Hard Times*, the scientific paradigm instructed people to 'Never wonder!' The new industrial division of time and the spread of communal clocks, individual time pieces, and the factory whistle ensured that even those who might have been attracted to wondering would have had little time to do so!

The utopian socialists articulated the romantics' yearning for 'wonder' as well as their rejection of the rigorous and inhuman delineation of time in the factory system. The utopian socialists also embodied the desire to overcome the divisions between personal aspirations and the goals and workings of social and political institutions. Pierson (1979: 17) notes that 'a radical severance between public and private spheres of life has been characteristic of all industrializing societies'. Those attracted to the early socialist movement were those who were especially susceptible to the social and psychological stresses which this severance caused.

The utopian socialists reacted to this disintegration of life with an attempt to create alternative cultural institutions. These ranged from cooperative stores to socialist Sunday schools, for example at Waihi and Christchurch. The socialist Sunday school children started their classes by declaring that they desired 'to be just and loving to all men and women, to work together as brother and sisters; to be kind to every living creature, and to help to form a new society with Justice for its foundation and love for its law' (Gustafson, 1980: 121). But the role of the alternative cultural institutions in the socialist movement was subsumed by the dominant trade union elements which sought tangible, material improvements for their members, rather than the creation of an alternative culture, or qualitatively different way of life.

The utopian socialists with their vision of life as a gorgeous feast made up of fine art, literature, beauty, free-thinking, and physical satisfaction could not compete with the mainstream socialist vision of improved material conditions for the mass of working people. But the romantic-inspired 'Back to Nature' movement in Britain in the late nineteenth century did have an important effect on the lives of many people. The romanticism about nature evident during this period, Gould (1988) argues, symbolized a deep-seated rejection of industrial development. Antipathy to the dominant trends of the age included a rejection of its 'objective spirit' and the intense nervous stimulation of urban life. By contrast, 'Back to Nature' carried pleasant connotations and conveyed notions of a simple and relaxed life, an alternative to industrial work, harmony with nature, the liberalization of social and sexual attitudes, and a sensitivity to animals, while promoting the creation of self-sufficient, self-governing communities where fact, value, thought, and feeling are equally valued and integrated (Gould, 1988: ix).

During this period, writers like Morris and Carpenter embraced socialism as part of their search for utopia. Their version of socialism, with its significant emotional and imaginative content, stood in contrast to the dominant Fabian emphasis on the control and management of the State to achieve socialist objectives. It was Fabian thought, rather than utopian socialism, which had the greatest influence on the subsequent evolution of the dominant Left, including the Labour parties in New Zealand and Britain. Early attempts to encourage attractive urban development in New Zealand through a Garden City Movement, for example, were criticized by a leading Labour activist, Bob Semple, as unimportant compared to the need to regulate capitalism for the benefit of all (Schrader, 1991: 25).

Qualities such as aesthetics and needs which reached beyond the material sphere received little attention from the dominant factions of the Left. Idealists who longed for a qualitatively different society, either in the early socialist movement or later among the advocates of the 'new politics' agenda, were not represented politically by the dominant Left party. Nor did this materialistic emphasis enable the traditional Left to later deal with its 'problem of the middle class', as working people and their offspring developed new goals and values along with their increased affluence. Offe (1984: 253), for example, claims that traditional democratic socialism is today being transformed into 'eco-socialism' (ecological socialism) as part of a process of overcoming the Left's fundamental problem of how to become more than a labour movement. The political Left is today in a tenuous position as it attempts to maintain the support of its traditional constituency on the one hand, at the same time as seeking to embrace aspects of the 'new politics' agenda—often promoted by the sons and daughters of its traditional constituency—on the other.

The Environmental Movement

Lest it be interpreted that this historical analysis of the emergence of green politics underestimates the importance of the environmental movement and concrete environmental issues, the relationship between these issues and the rise of green politics needs to be made explicit. Changing social patterns and values can only explain part of the rise of green politics. Concrete issues have catalysed the green movement around the world (Rüdig and Lowe, 1986). This relationship was particularly important in New Zealand, where 'wilderness' issues were the main focus of the environmental movement (Rainbow, 1992). The Save Manapouri Campaign, the Native Forest Action Council, and campaigns against the Clyde Dam and a second aluminium smelter at Aramoana near Dunedin were all examples of movements which gained high public profiles and helped to create a green constituency in New Zealand politics (Wilson, 1982).

While institutional mechanisms have been developed in response to international environmental pressures and issues, the sense of environmental crisis has only worsened since the environment first became a political issue. New Zealand's representation at the 1972 Stockholm Conference on the Environment led to the formation of a Commission for the Environment that same year, and New Zealand participated in the World Population Conference in 1974. But in spite of the institutional recognition of the environment, a range of non-governmental organizations and increased media attention have raised the profile of the global environmental crisis, regularly bringing new issues and examples of despoliation to the fore. Toxic wastes, industrial pollution, nuclear accidents, acid rain, global warming, and climatic change, the unprecedented extinction of species, massive deforestation, a reduction in biodiversity, and uncontrolled population growth: a plethora of environmental ills demand responses and provide motivation for green politics around the world. There is no single green response to the environmental crisis, however, and an analysis of the variety of green responses forms an important part of this book. Suffice to say that the 'apocalyptic tone' (Dobson, 1990: 22) prevalent in green politics since the 'limits to growth' genre of literature of the 1970s, while galvanizing people to action, is no basis for a constructive long-term strategy to seriously tackle the eco-crisis. Nor can the environment be addressed without reference to social, political, and economic issues. An awareness of the interconnected nature of the crises of modern society leads most greens to 'desire to restructure the whole of political, social and economic life' (Dobson, 1990: 3) in the same way that their utopian socialist, New Left, and new social movement forebears sought, and seek still, to create a qualitatively different way of life.

Conclusion

This list of the origins and antecedents of green politics is not exhaustive by any means. There are, for example, those who assert that green politics is the logical successor to the anarchist tradition (Leach, 1988). Others argue that green politics is an unprecedented phenomenon inspired by discoveries in science which point towards the likelihood of a revolutionary new paradigm. They claim that this 'new paradigm' and its green political manifestation supersede traditional, industrially based political divisions (Spretnak and Capra, 1986).

What I have attempted to demonstrate above, however, is that, whether in its concerns with internal party processes or in its critical attitudes towards the costs to the environment and quality of life from industrial development, the concerns of contemporary green politics have been articulated by other movements throughout the modern era. Green politics is not ahistorical. There have been a series of critical responses to industrialism and modernity, amongst which factions of the early socialist movement were significant. Green politics can usefully be seen as linked to the utopian socialist tradition, a tradition which embraced an enlightened attitude to social issues *as well as* to 'the environment'. There is a remarkable degree of continuity between the utopian socialist and contemporary green movements, in spite of the century which separates them. This suggests, importantly, that green politics is more than a passing phase, an ephemeral issue with no connections to the past and unlikely to endure into the future.

To locate green politics on a historical continuum of anti-modernist movements is to contest the claims of those who would assert that green politics is purely an environmental movement or a movement based on an ecological determinism. Environmental concerns themselves are no guarantee of a political programme, and a concern solely with environmental issues can lead just as easily to a totalitarian political future as it can to a pluralist, democratic one. There is a real need, then, for green politics to acknowledge historical antecedents in order that its concern with the environmental crisis does not result in connected issues of a social and political nature being ignored.

It is impossible to attempt to solve environmental issues without addressing questions of social and political organization and the values which underlie them. According to the discipline of ecology which influences much green thought, everything is interconnected. It is entirely consistent with the greens' own holistic and ecological intellectual foundations to claim that the environment cannot be separated from social, political, and economic questions. By placing environmentalism and green politics in a broader context, it challenges the schemes of those who would demean not only democracy but even 'civilization' in their commitment to the potentially authoritarian implications of a concern solely with the environment.

A historical approach suggests that green politics can learn from its antecedents and develop tactics and strategies which will ensure greater success in advancing towards green objectives. If green politics is not to be just another minority sect preaching moralism against a dominant materialist system, considerable thought needs to be given as to how green objectives are to be achieved. This requires, for example, that green politics enhance its role as a policy generator. All of the critical global problems of our times demand very practical policies at a local, regional, or national level.

Effective green strategies will be more easily generated if we understand the historical impulses which influence green attitudes to issues such as internal party organization and political power. Most latterly, the New Left and new social movement attitudes towards hierarchy and participation have had a deep and debilitating influence on green political structures. Understanding the historical background to the greens' anarchic impulses is one way of ensuring that they are able to be superseded by more effective strategies for achieving change based less on the emotional dissonance felt by those in the green movement than on a rational commitment to finding workable solutions to the very real problems which greens have been at the forefront of bringing to public attention.

Current examples of reform politics, epitomized by labour and social democratic parties, have abandoned their positions at the cutting edge of social change. This has left a political vacuum in contemporary politics which has been filled by the greens. Green politics speaks for those who, like its utopian and romantic forebears, feel alienated from the dominant values of the age. The greens represent those who feel that too much destruction has been done in the name of a 'progress' with which they feel little affinity. The challenge is to turn this response into a hard-headed programme for reform.

2   Green Politics as Holistic Politics

Unity is never absent from us, but seldom realised. (Phipps, 1990:108)

In spite of the existence of a succession of anti-modernist movements for more than a century, most people do not have a clear picture of how a green society might look. In the public mind, there is little understanding of green politics beyond an awareness of its concern for 'the environment'. Furthermore, the effects of many aspects of green politics in practice have not been thought through fully. Nor is there a concrete programme for the realization of a green society. An important aspect of any political ideology—its strategy for achieving its objectives—is underdeveloped in green politics.

The role of the State and issues such as private versus public ownership are the basis of most conventional analyses of political ideologies. Political differences are usually located along a political spectrum ranging from 'Left' (socialism) to 'Right' (conservatism). It is not always easy to locate green politics on this conventional political continuum, however. A range of issues—from environmental protection to access to abortion—have proved difficult to position on the Left–Right spectrum. As a consequence, alternative models have been suggested, such as Vedung's (1989) spectrum with 'ecology/environment' at one end and 'development' at the other. Green politics is clearly placed at the ecology/environment end of such a spectrum, in a position consistently shared by no other significant contemporary political movement in modern industrial societies. The traditional Left–Right spectrum assumes a shared commitment to development and 'progress'. By contrast, the green ideology takes a largely critical view of 'progress' as it is defined in contemporary industrial society.

Taken to its logical conclusion, green politics poses a fundamental challenge to the development-industrial ethic which has dominated Western societies during the modern era. This is not just because of the nature of the issues which green politics brings to political agendas, but also because of the 'classical' definition of politics which green politics embraces. This definition is based on the premise that politics takes place at all levels of society, not just at the level of the State, which is the typical focus of modern politics. With a 'classical' definition of politics the personal becomes political as individual happiness and the composition of the 'good life' become explicit subjects for political attention. This definition disregards the limits which liberal democracy places on political jurisdiction over a sphere of inviolable individual rights, challenging the 'sharp delineation of the state's powers' (Fukuyama, 1992: 15) which liberal democracy entails.

Green politics explicitly addresses the areas of personal behaviour and values, as well as what we have come to regard as 'politics' in the conventional sense. Green politics consequently often acts as an ethical framework for guiding the daily behaviour of its adherents, not limiting its influence merely to the realm of public policy and the electoral sphere. Green politics has potentially religious overtones as it provides a set of values and standards by which people might lead their daily lives.

The all-encompassing nature of green politics is reflected in the assertion that the new movements—like the greens—speak for those longing for a 'softer society' (Inglehart in Friberg and Hettne, 1985: 219), or that, like the New Left, they impart 'political dignity to the tenderer emotions' (Roszak in Caute, 1989: 50). A world of beauty, institutions with a human face, an abhorrence of waste, authentic human relationships: these are the concerns which motivate many adherents of green politics. Such emotions are hard to quantify politically, but cannot be ignored as potent sources of political inspiration. Together, these kinds of desires represent the hope that a better society is possible without destroying the past or sacrificing the rights and resources of current and future generations. Green politics focuses on saving the environment from the negative effects of development, on bringing technology back within human control, and on promoting human concerns against the pervasiveness of economic values. But, above all, green politics represents the desire for a coherent world-view with which to understand contemporary society and to guide its choices.

> Many Green Party members come from ecology, peace, feminism, or other political strands because they've been through these things and want a more holistic analysis and a philosophy that ties it all together.(Dann in McVarish, 1982: 89)

Green politics reflects a renewed search for 'wholeness', a desire to 'bring together again our divided existence, to repair the disintegration, and replace it with a consciousness of totality' (Hülsberg, 1989: 137). The greens, like their romantic forebears, seek a holistic world-view which recognizes the importance of relationships and interconnectedness: 'the pattern which connects the crab to the lobster and the orchid to the primrose and all four to me' (Maren-Griseback in Spretnak and Capra, 1986: 33). Greens seek to apply this holistic paradigm to the political realm, including to their own intra-party processes. Green politics attempts a new way of 'doing politics', placing considerable emphasis on the importance of the 'process' by which decisions are made. This reflects a rejection of modern political processes and seeks to set new standards of political behaviour through open, participatory structures and accountable representatives.

Key components of the holistic green ideology include:
- the internal practices of green politics should be consistent with the kind of vision greens advocate for society generally
- waste, especially of non-renewable resources, is immoral

- values which countenance the expenditure of vast sums on unnecessary activities like the arms race, gimmicky products, and advertising, while millions of people lack basic amenities and the environment cries out for massive reparations, are repugnant
- politics should include issues of personal change and should embrace basic humanist principles about the right of all people to develop their abilities to the fullest, to be fulfilled, and to enjoy authentic relationships
- the hegemony of money values and materialism must be rejected
- aesthetic concerns, particularly for 'a world of beauty' which exists in harmony with the natural environment, are crucial
- all issues are interconnected, meaning the need for, among other things, a global approach to policy-making
- modern society is unbalanced, putting too much faith in science, technology, and 'specialists', while devaluing folk wisdom, and qualities such as intuition, compassion, altruism, and spirituality
- human activities must no longer treat the natural environment as expendable—they must become 'sustainable', environmentally and socially.

Together, the components of this holistic agenda comprise a substantial critique of modern society.

The Rejection of Modernity

At the core of green politics is a rejection of modernity; that is, the unique form of ('modern') society since the seventeenth century. Modernity corresponds to the period of industrial society. Its central components are the bêtes noires of the green movement: industrialism and bureaucracy. The green critique of industrialism extends beyond concern with the adverse environmental effects arising from a particular mode of production. For industrialism has been described as a complete way of life, 'a culture, a set of social institutions, an epistemology, an integrated way of life' (Toffler, 1983: 87). Industrialization has involved the use of new energy sources, new raw materials, and new inventions, especially in the fields of transport and communications. It has involved the application of new techniques, enhanced roles for science and technology, and the creation of the 'factory system'.

Industrialism has also had a marked effect on the individual's life. It has not only provided an unprecedented array of goods, it has also promoted the values of individual egoism, competitiveness, and acquisitiveness. Because all efforts are concentrated on working for the productive process, the individual is encouraged 'not to speculate on the past, develop as an individual, or contemplate the future' (Gould, 1988: 163). Industrialism has led to a speeding up of time which many individuals experience on a subjective level. Because of mass production, industrialism has encouraged standardized patterns of living and consumption, encouraging uniformity while discouraging individuality, spontaneity, and non-material values. All other values tend to be subsumed by the economic imperative, and bureaucratic relationships replace long-established personal bonds.

Hand in hand with industrialism and the rise of the modern State has been the ascendancy of centralization and what Weber (in Pakulski, 1990) describes as 'formal rationalism', which is characterized by bureaucratic organization. Bureaucracy 'epitomise[s] the modern social order with its intrusive regulations, formalism, centralization and, above all, instrumental, calculable and value-neutral mode of operation' (Pakulski, 1990: 160). Rationality in this context refers to social arrangements which are deliberate, systematic, and impersonal. This leads to an increasingly 'disenchanted' world-view shaped by scientists, industrialists, and bureaucrats. Importantly, this kind of rationality works against relationships built on altruism, love, compassion, and loyalty. Traditional relationships are replaced with increased freedom from tradition, and increasing susceptibility to the anonymous and impersonal forces of the market. Above all, as Weber (in Pakulski, 1990) accurately predicted, this ascendancy of formal rationality would lead to the domination of politics by bureaucratic parties and the State.

A critical attitude towards modernity and its component parts unites the diverse strands of green politics. Green politics represents 'a revolt against the formalism and centralism inherent in bureaucratised politics, the institutions of the state, the market and modern mass culture' (Pakulski, 1990: 172). Empirical research of green activists cited by Pakulski (1990) shows that specific causes such as environmental issues generate concern not just because of their own intrinsic importance, but because of *'the general problems these issues illustrate and signify'* (Pakulski, 1990: 170—my emphasis).

The response to the perception of a general malfunctioning of modern society is the hope that green politics represents a new stage in social evolution, from industrial society to a new stage of human development variously described as postindustrial, postmodern, and even, by Boulding (1978: 136), 'post-civilisation'. Each of these labels speaks of the prospect of a fundamental transformation of modern society similar in magnitude to that which occurred in the transition from agrarian to industrial society. Greens believe that the next stage of social development will lead to a 'systems break', with bureaucracy and industrialism being superseded by a new holistic paradigm based on overcoming the separation of humans from nature.

The Differences Within Green Politics

In spite of a shared rejection of 'modernity' and the values upon which it is based, there is no uniform green response to the crises of modern society. Green politics is not an undifferentiated whole, which makes it difficult to speak of 'the green ideology'. As with most political ideologies, within green politics there are differences around issues such as the pace and means of change and the relative importance of environmental versus social issues. These differences have been described in a variety of ways. Some writers have identified the divisions within green politics in traditional Left–Right terms (Hülsberg, 1988). It is most common, however, to identify a light–dark spectrum within green politics. The lighter end of the spectrum supports reform through existing political avenues and tends to seek a balance between social and environmental concerns. Those who desire 'fundamental' change, namely the dissolution of industrial society, appear at the spectrum's darker end and tend to be motivated by a sense of impending ecological catastrophe. These divisions parallel the splits between the reformers and revolutionaries in the early socialist movement (Papadakis, 1989; Porritt and Winner, 1989).

The assertion that green politics is essentially an anti-modernist movement provides a useful starting point for explaining the differences along the light–dark spectrum. Responses to anti-modernism can take a variety of forms, but might be categorized between those who seek to return to the past (pre-modernists) and those who seek a qualitatively different kind of future (postmodernists). Berger *et al.* (1973: 188) describe the various 'counter-formations' proposed to replace existing social arrangements as 'reactionary' or 'revolutionary', 'depending on whether the resolution of the discontents has been sought in the past or the future'. These categories may also be useful for understanding the spectrum within green politics.

The light or dark orientation of green members affects the nature of the respective green party. Müller-Rommel (1985) and Rüdig (1985), not untypically, assert that there are two different kinds of green parties: 'pure' environment parties (dark green), and 'rainbow-coalition' (light green) type parties which embrace a wide range of movements and address social as well as environmental concerns. In fact, elements of both these party types exist within most green parties, often coexisting alongside each other with considerable tension.

Party orientations also change over time. While members may initially join primarily out of concern with environmental issues, political involvement can increase the awareness of the interconnectedness of environmental issues with most other aspects of social and economic policy. The nascent Values Party, for example, was initially formed around concerns about consumerism, population growth, and economic growth. While it was easy for Values to unite around this generalized policy for the 1972 and 1975 elections, when the process of developing detailed policies began after 1975 considerable intra-party conflict was generated. By 1978, a sizeable group within the party had moved beyond vague idealism to embrace a set of policies designed to explicitly challenge private ownership of land and capital. Not all party members agreed with the direction of this policy development, but it did reflect the desire of some in the party hierarchy to move beyond being labelled as ' "airy-fairy" idealists' (*Evening Post*, 16 May 1978) and to develop a full raft of detailed policies befitting a serious political party.

Parties do shift from 'purposive', or strongly ideological positions, to 'pragmatic' perspectives over time. Issue-oriented parties usually find the need to address issues other than those with which they are primarily concerned, if they are to win significant public support. The Values Party's earlier idealistic manifestos, for example, gave way to a publication in 1978 with a cover featuring a protesting worker holding a placard reading 'Make My Dollar Go Further'. In a similar attempt to broaden their appeal, the German Greens put their environmental policy at the end of their 1987 programme (Papadakis, 1989: 62). But these populist gestures did not go unopposed within the respective parties, with fundamentalists decrying any attempt to dilute the challenge of green politics to industrial society.

The German Greens (Die Grünen) provide a conspicuous example of typical intra-party divisions. They were racked with party splits during the 1980s, and it appears that the loss of representation in the Bundestag in 1990 has only served to exacerbate these differences. The key splits were between the (dark green) fundamentalists, who were committed to a total transformation of industrial society, and those (light greens) labelled 'realists', who accepted the need to work within existing institutions to achieve reforms.

Rudolf Bahro was a leading fundamentalist within the German Greens until his departure from the Party over the issue of animal rights in 1986. He argued (Bahro, 1986: 218) that the Greens' central task was to stop the process of industrial society, '. . . not to create a space for minorities, but to create a new solution for the whole of society'. Any compromise with existing institutions in the form of political arrangements with established political parties, for example, simply gives a new lease of life to the destructive industrial system. Because the arms race and the destruction of the environment is a direct result of the industrial juggernaut, the greens' role is to assist the disintegration of the industrial world. Greens must contribute to the creation of a qualitatively different way of life based on a new consciousness and a new psychology. Bahro (1986: 90) called for 'a new Benedictine order' and for the creation of communal living on a widespread basis. Fundamentalists argue that continued 'progress' is unsustainable, environmentally, economically, and psychologically. The era of modernity has been an historical aberration which reflects the victory of unbalanced rationality and an instrumental view of the world based on attributing worth to nature only insofar as it is of use-value to humans.

Green fundamentalists believe that civilization needs to be 'unmade', that the natural world stands as a check against the manipulative power of the human-created society. Wilderness becomes a reference point against which the artificiality of modern life can be gauged because:

> Radical environmentalism takes on a significance beyond its aggressive call for the protection of nature and embraces a vision of human freedom unacceptable to civilization, with its urge to contain, repress, and inscribe all that is wild. (Manes, 1990: 221)

Civilization itself is identified as the key problem, with fundamentalist greens often calling for a return to hunter-gatherer societies or the emulation of the lives of tribal peoples. Technology is rejected in this world-view because it places undue emphasis on the notion of utility, and threatens to control every aspect of human life. Economic activity is a measure of human exploitation of the earth, and industrialism is perceived as the ultimate means of degrading the planet and the intrinsically valuable species it supports.

In contrast to the fundamentalist ideologues are those whom Kitschelt (1988: 130–1) describes as lobbyists and pragmatists. These greens have a less dogmatic idea of the shape of future society, are more tolerant of diversity, and possess a respect for democracy and freedom frequently lacking in their fundamentalist colleagues. The 'realists' are less attracted to the notion of the greens as a 'permanent opposition' in industrial societies, rejecting the strategic bankruptcy, the self-righteousness, and the lack of opportunities to positively contribute to the shape of future society which such a stance entails. They are not motivated by utopianism, as the former Green Environment Minister in the German state of Hessen, Joschka Fischer, articulates:

> I am no longer motivated by utopias, but by the description of existing conditions. The ecological crisis, the arms race, the rise in criminality—those are more than enough for me. . . . If we can take one step in the right direction, one step which moves us away from the abyss, then there is sufficient justification for the existence of the [Green] party. (Hülsberg, 1988: 128)

The realists rightly point to the fact that there is no strategy for the realization of a qualitatively different kind of society as advocated by the fundamentalists, and that even small-scale reforms are difficult to achieve. They advocate strategies such as working cooperatively with other parties, and particularly with sympathetic elements on the Left of established labour and social democratic parties. In several countries, green parties, because they hold the balance of power, have enabled the established Left parties to govern in exchange for the implementation of certain policies or positions of influence. Developing credible green policies will ensure that the established Left parties have to keep looking leftwards to address the issues and proposals which the greens raise.

Tasmania has recently provided such an example with Greens and the Australian Labor Party (ALP) cooperating within the Tasmanian Parliament. Although the arrangement survived for only a limited period, it was notable for several tangible policy achievements for the Greens. Similar successes have occurred as a result of coalitions of the Greens and Social Democrats in certain states and cities in Germany, but examples of greens working with conservative or right-wing parties are rare. In New Zealand, the Green Party has joined an alliance of minor parties with a leftist orientation, but it is difficult to ascertain whether this reflects a shared ideological approach among the Alliance parties, or whether it is an exercise in short-term electoral expediency in the face of an electoral system biased against minor parties.

The Green Emphasis on Cultural Change

Light and dark greens share assumptions about the need for radical 'cultural change' if modern society is to be diverted from its current destructive course. Green theories of change differ markedly from conventional theories of social transformation. Theories of social and political change during the last century have usually employed Marxism as their benchmark. The traditional Marxist view is that society and its institutions reflect the mode of production. The new movements believe that this economic-determinist analysis denies the complexity of the cultural sphere because they believe that *ideas and values* are central determinants of the kind of society we live in. Put simply, the mode of production is itself the product of a certain set of ideas and values, rather than vice versa. Therefore, the areas of values and ideas have to be addressed, for they provide the cultural foundations of particular forms of social organization. Touraine (Jennett and Stewart, 1989: 2) argues that in postindustrial society, the transformative struggle will take place at the cultural level, and that the conflict embodied by the new movements is about the control of the dominant cultural patterns in society. Cultural patterns, like paradigms, govern the norms of society which, in turn, play a central role in determining the behaviour of individuals and institutions, as well as the mode of production.

To make the connection between values and current social problems explicit, greens subscribe to the view that 'the worldwide ecological crisis is fundamentally a crisis of mind and spirit':

> . . . the crisis of ecological scarcity can be viewed as primarily a moral crisis in which the ugliness and destruction outside us in our environment simply mirrors the spiritual wasteland within; the sickness of the earth reflects the sickness in the soul of modern industrial man, whose life is given over to gain, to the disease of endless getting and spending that can never satisfy his deeper aspirations and must eventually end in cultural, spiritual and physical death. (Ophuls, 1977: 231)

If what happens in the world is a reflection of people's inner worlds, then people's thoughts and feelings become political. Forms of politics which have included the individual and cultural sphere have traditionally been associated with totalitarian politics. A good example is the attempt of the Bolsheviks to create a 'New Soviet Man' who would embody all those characteristics considered desirable by those wanting to build communist society. This can be contrasted to conventional politics in liberal democratic societies which is usually concerned with changes at the level of the State, and which delineates clearly between 'public' and 'private' worlds. But, according to the green perspective, if real change is to occur, then we must look beyond conventional changes of guard at the level of the State and begin to change the way people think and what they value, i.e., the dominant cultural patterns in society must be transformed. Green politics fulfils Berman's (1981: 23) prediction that: 'Some types of holistic . . . consciousness and a corresponding socio-political formation, have to emerge if we are to survive as a species'.

Greens tend to the view that modern society is no longer functioning properly, not only because of obvious issues like the environmental crisis, but because 'our moral and spiritual abilities have not developed as fast as our technical ones' (Phipps, 1990: 99). Green politics reasserts the need for ethics to guide our technological and political choices. Both capitalism and its Marxist alternatives have proved inadequate because of their preoccupation with the material world, and their neglect of important areas of life such as emotions and spirituality. If the current malaise in modern society is a reflection of people's psyches, change will not occur simply by altering institutions or the mode of production. It becomes necessary to work for 'the transformation of the spirit, without which, any alteration of institutions is doomed to failure' (Buber, 1958: xv).

The New Paradigm

> Modern culture—as we all recognise since we live in the belly of the beast—is based on mechanistic analysis and control of human systems as well as Nature, rootless cosmopolitanism, nationalistic chauvinism, sterile secularism, and monoculture shaped by mass media. (Spretnak, 1986: 29)

A paradigm is a set of beliefs, or a framework with which the world is comprehended and made sense of. Some writers argue that green politics is the political manifestation of a totally new paradigm. The 'new paradigm' is allegedly responsible for the panoply of changes which are now occurring at every level of society. The idea that the existing paradigm is in the process of being superseded has been popularized by the concepts of a 'New Age', 'The Aquarian Age', Galbraith's (1977) 'Age of Uncertainty', Drucker's (1969) 'Age of Discontinuity', and Toffler's (1980) 'Third Wave'. Each of these concepts, and many others, suggest that society is undergoing a fundamental transformation to a different set of values and behaviours. Eisler (1988), for example, argues that society is being transformed from a dominator to a partnership culture, reflecting a shift from male to female values: 'The ferment of our modern age as a time of unprecedented technological change is providing the opportunity for social change—and potentially for a fundamental social transformation' (Eisler, 1988: 170). *Megatrends* author, John Naisbitt (1982), identifies recurring trends by which this new age can be identified. It includes, he suggests, factors such as the growth of a global economy, the reality of an information society, the growth in self-care, increased personal choices, the breakdown of hierarchies, and the spread of decentralization.

Such social and political developments are underscored by a new paradigm which is gaining legitimacy because of the growing visibility of the contradictions arising out of the current rational-scientific paradigm. The new paradigm legitimizes a different set of values, a different world-view, and new cultural patterns. The problems which modern society faces, proponents of the new paradigm assert, are the results of a singular crisis of misperception resulting from the existing paradigm (Spretnak, 1986: 19). The new paradigm is holistic, in contrast to the current dominant paradigm which has legitimized the separation of humans and nature upon which industrialism has depended. The existing paradigm replaced a world of superstition and religion with the conception of a world which could be known by mechanistic rules and surety. The world was akin to a machine which, like people and other organisms, could be understood by reducing it to its component parts, each of which was governed by a set of knowable laws.

By contrast, the new paradigm means that 'the universe is no longer seen as a machine, made up of a multitude of objects, but has to be pictured as one indivisible dynamic whole whose parts are essentially interrelated and can be understood only as patterns of a cosmic process' (Capra in Hutton, 1987: 21). The new paradigm has been heavily influenced by the discoveries in the scientific field of New Physics which asserts, among other things, the inter-connectedness of all phenomena, the constancy of change, and the inevitable subjectivity of the 'scientist'.

The new paradigm has also been informed by the science of ecology which focuses on the interrelationship between seemingly different phenomena. Spretnak's and Capra's seminal work on green politics (1986) quotes philosophy professor and former German Green Party spokesperson Maren-Griseback's assertion that 'ecology is the secure and scientifically sound foundation for the entire Green philosophy' (Spretnak and Capra, 1986: 32). Ecology has the respectability of being a 'science' which legitimizes the search for what is essentially a metaphysical search for wholeness or holism. Ecology has a respectability, an academic legitimacy which the romantic belief in the poetic notion that 'indissoluble ties unite all things' (Gould, 1988: 18) did not enjoy. The desired outcome for both ecology and romanticism is the same: legitimacy for a framework which would confirm the intuitive belief in the interconnectedness of all life. Capra (1982) argues that green politics is the political manifestation of this new knowledge about the interconnectedness of all life on earth.

Writers like Capra (1982) and Ferguson (1980) draw heavily on scientific innovations to support their case for the validity of a new holistic paradigm, of which green politics is the political expression. Conclusions drawn from chemical reactions in which there were sudden and spontaneous reorganizations have been applied to the social sphere with all the certainty of a Marxist prediction about the inevitable shape of historical developments. But, as Anderson (1990: 246) argues: 'It is a long, long leap from describing something that happens in a chemical reaction to forecasting global cultural change'. The existence or otherwise of the 'new paradigm' remains contentious. What can be safely stated is that the desire to legitimize a new paradigm is part of a broader struggle, of which green politics is but one part, to bring new perspectives to bear on the crises of modern society and to validate new patterns of development in the future. The new paradigm seeks to legitimize a world-view which is holistic and which particularly asserts the necessity of overcoming the separation of humans from the natural environment.

The Revolt of the Metaphysical

Even those who have a less critical view of 'progress' than greens have recognized the importance of spiritual and metaphysical issues to the future shape of our society. Nisbet's *History of the Idea of Progress* acknowledges:

> Only . . . in the context of a true culture in which the core is a deep and wide sense of the *sacred* are we likely to regain the vital conditions of progress itself— past, present, and future. (Nisbet, 1980: 357)

The kind of society we envisage is inextricably linked with our answers to basic metaphysical questions about the meaning and purpose of life. The greens represent the revolt not just of the social and political, but also of the metaphysical, as they pose afresh basic questions about the purpose of life in modern societies.

The metaphysical in green politics is connected to the calls for a re-enchantment of the world and a renewed awareness of the 'mystery of creation', or what Albert Schweitzer called the 'Reverence for Life'. Schweitzer deeply influenced people like Rachel Carson (pioneering American ecologist) with his assertion that: 'we are not truly civilized if we concern ourselves only with the relation of man to man. What is important is the relation of man to all life' (Carson in Brooks, 1989: 316). Schumacher, a prominent green philosopher, called for a 'metaphysical reconstruction' because 'a person who sees the world as being in some sense sacred would be morally, psychologically, and politically incapable of allowing such an infinitely precious world to be destroyed' (Phipps, 1990: 43). Instead, scientific rationalism has disposed of any concept of the sacred, of the enchanted: 'reality has become dreary, flat and utilitarian, leaving a great void in the souls of men which they seek to fill by furious activity and through various devices and substitutes' (Bell, 1982: 502).

These kinds of concerns are reflected within the green movement by the philosophical framework of 'deep ecology', a philosophy which rejects anthropocentrism and any suggestion of a hierarchy of species. Deep ecology rejects the assumption that all life forms are simply in existence to serve human ends. It attributes nature with certain intrinsic qualities which are quite separate from any use-value for humans. Deep ecologists assert the need for humans to develop a sense of relatedness with all species; an awareness that to harm one species is to harm all species. The growth of animal rights movements and of vegetarianism demonstrates the potency of these views in contemporary society.

The Personal and Psychological Significance of Green Politics

The personal significance of green politics to its activists and adherents should not be underestimated. Green politics has been described as a 'form of ethical life' closely resembling a secular religion (Leach, 1988: 5). Green politics is more likely than conventional political allegiances to influence the daily behaviour of its adherents, determining decisions about what, if any, products are bought, what forms of transport are used, and what ways people will live, including how many children they might have. Spretnak (1986: 45) asserts that 'Green politics is about values in our daily lives, how we live and work and play'. This is why the public arena of adversarial politics, which is usually the central focus of established parties' attention, is not the sole focus of green political attention.

Perhaps nothing indicates so clearly the uniqueness of green politics in contemporary politics than its willingness to explicitly address spiritual and metaphysical issues. The blatant misuse of spirituality and religion by regimes like the German Nazis has inhibited political attempts at addressing matters of the spirit in modern society. But many green writings refer to spiritual matters. The Values Party's 1972 manifesto claimed, for example, that New Zealand was in the grip of a new depression: 'It is a despair which comes usually with physical poverty. But New Zealand's peculiar malady is not physical poverty; it is spiritual poverty' (Values Party, 1972: 1).

Charlene Spretnak, author of *The Spiritual Dimension of Green Politics*, argues that 'we can't solve our political problems without explicitly addressing our spiritual ones' (Spretnak, 1988: 18). Many issues of personal and social import are influenced by matters of the spirit. The compulsive material consumption which greens find an anathema can be explained, for example, as the result of an inner emptiness or spiritual void which people attempt to appease by the relentless pursuit of material goods. The dominance of materialism and the concentration on worldly goods becomes possible 'because there is no inner life in modern, technological society' (Spretnak, 1988: 16). Modern society, Heller and Feher (1988: 14) assert, is epitomized by universal dissatisfaction, leading to an endless cycle of unstoppable growth because 'without dissatisfaction modern society could no longer reproduce itself'. A highly ranked General Motors official argued that business needs to create a dissatisfied customer; its mission is the 'organized creation of dissatisfaction' (Kettering in Schor, 1991: 120). Face-lifted vehicles every year are one of the outcomes of this plan to ensure that dissatisfaction is constantly fuelled. In this way, the acquisitiveness and materialism which greens reject is explicitly connected to broader psychological and metaphysical issues.

Recognizing the importance of spirituality in a satisfying life, many New Zealanders in recent years have turned to different spiritual paths. Spiritual gurus and New Age speakers attract large crowds to their gatherings, which suggests that a significant number of people are embarking on various forms of spiritual searching. Because green politics reflects the desire for a politics which is a part of life, rather than separate from it, it is not unreasonable to assert that green politics is a political manifestation of that same search for meaning and ethics. This is one of the reasons that a survey of New Zealand Green Party conference goers in 1990 revealed an interesting array of spiritual preferences, ranging from 'Gaians'—worshippers of Gaia, or Earth—to 'post-Christians', Pagans, and Pantheists. This was in marked contrast to the (more conventional) religions mentioned by conference goers from the mainstream parties surveyed in the same year (Miller, 1991: 64).

Utopianism

Green politics is the most obvious form of utopian politics in contemporary politics. Utopianism emerges at a time when severe social dislocation and the breakdown of traditional values leads to the call for new norms relevant to a new era. Utopian socialism emerged in response to the dislocations caused by rapid industrialization. Green politics reflects the same response to the increasingly conspicuous effects of industrialization on the natural environment and on the quality of life in modern industrial societies.

Utopianism has a special relevance to New Zealand, and green politics may well be tapping into a recurring utopian streak in New Zealand politics (Cleveland, 1979). New Zealand's isolation and the sense in which it has provided a haven for refugees from the despoiled northern hemisphere have fuelled this utopianism. Many of the 'environmental refugees' who have made New Zealand their home in recent decades have shared the same sense of New Zealand as a place to escape the worst excesses of northern hemisphere industrialism as the first European settlers. Utopianism 'plays a central role in New Zealand's perception of itself' (Sargent in McLeod, 1992: 26) and, as mentioned earlier, New Zealand is one of the few Western countries in the world to have given State support to utopian ideas (namely the ohu supported by the Third Labour Government). For early settlers New Zealand was an ideal place to depart from established social patterns, and the idea that British society could be re-established with all the bad bits removed is, according to Sargent (in McLeod, 1992: 26), hard to surpass as an example of a utopian aspiration.

Ironically, however, antipodean utopianism has, in part, been responsible for some of the environmental degradation in New Zealand, at least until the 1970s. A strong component of the Labour Party's 'scientific utopianism' after 1938 was the belief in progress by superior technology (Cleveland, 1979: 12). The socialist vision of reshaping the environment through human effort fitted very comfortably with, for example, the harnessing of waterways for hydroelectric development. As in Tasmania, the New Zealand utopian vision included the idea that the wilderness was not there to oppress, but rather to be harnessed to enable the 'good life': 'which meant the good society: high employment and taxes sufficient to guarantee an adequate system of public health, welfare and education' (Lohrey, 1990: 94). Environmental destruction was an integral component of antipodean utopianism until the 1970s when a new utopian vision came to the fore. At this point, the nascent green movement began to present a view of New Zealand's isolation and marginalization as a godsend in a world increasingly faced with pollution and the threat of nuclear holocaust.

In the past, it was possible—at least in a European context—to talk of mass migration to a potential utopia: the North American continent, Australia, or New Zealand. More latterly, science fiction has fantasized about the prospect of human habitation of outer space. Now, in keeping with the general green emphasis on limiting the adverse effects of human behaviour on the external world, there are calls to redefine frontiers, from external to internal. Bahro (1984: 221) argues that: 'I believe we have now reached the point where humanity has to find a new stable life-form in which its forward development is an internal journey rather than an external expansion'. Humanity cannot continue destroying its hinterlands; future growth will have to take place within each individual rather than across new physical frontiers.

Peace, between individuals and nations, as well as between people and the natural environment, is an enduring utopian goal. The 1975 Values Party manifesto stated: 'Our goal is a new age in which community is more important than materialism and man learns to live in harmony with the rest of nature rather than against it' (Values Party, 1975: 78). Living in harmony with other people and with Nature is based on the premise that all people, as well as people and the natural world (including other species), are inextricably linked and interdependent. The modern peace movement is a contemporary expression of the utopian hope that people might live together in harmony. The peace movement, particularly in a European context, was one of the major constituencies for the fledgling green parties in the 1980s. The peace movement in New Zealand worked particularly hard to establish a network of 'nuclear-free zones' throughout New Zealand from the mid-1970s. The nuclear-free issue formed an integral part of the green agenda in New Zealand throughout the 1980s, with the Labour Party successfully able to capitalize on this issue. It gained the support of the latent green constituency at least in the 1984 and 1987 elections because of its nuclear-free policies.

The Green Attitude Towards Organization

The greens' holistic ideology influences not just the green openness to spirituality and its utopian aspirations, but also shapes green approaches to internal party organization. A holistic approach demands that political means cannot be separated from the desired political ends. Much attention is paid within green parties to the question of political 'process', and party structures are designed to 'prefigure' the kind of society which the greens aspire to create. A 'small is beautiful' ideology pervades the green attitude to organizations because, it is claimed, 'social organization is most compatible with nature when it is small-scale and based on principles of direct democracy' (Hutton, 1987: 18). The green attitudes to organization have often, however, made green party practices less than successful in their dealings with conventional political institutions. The 'rotation' of party personnel and other attempts to prevent the emergence of a single leader have often made it hard for the public to identify particular figures with green politics. Internal green party practices are often fraught with tension due to unclear responsibilities and non-hierarchicalism being taken to extremes, so that green parties have been described as 'organized anarchies' (Andersen, 1990: 110). The attitudes of greens to organization, as well as to politics and power generally, are a result of the influence and experiences of the counter-culture, new social movements, and New Left, all of which embraced 'a particular set of libertarian attitudes to power, organization, democracy, economics and culture' (Landry *et al.*, 1985: 3).

Green politics frequently asserts that decentralization is both desirable and necessary for all political institutions, from internal party processes to the State. It is claimed that if people make the basic determining decisions which affect their own lives and communities, then their self-confidence increases, engendering a sense of personal efficacy as well as empowering communities. Because decisions will be made by those most affected, and the fact that the people who make the decisions will have to live with the outcomes, local decisions will be made with greater responsibility than those made by people not directly affected. Decentralization addresses the issue of an alienation from unresponsive, centralized political institutions, and challenges the concentration of power and wealth by a political and economic élite who largely ignore the wishes of the majority of people. Decentralization and more participatory forms of democracy reflect the desire for an 'emancipation from authority'. It is also seen as essential to facilitating the new channels of participation sought in order to express a range of new issues ignored by mainstream politics.

Green politics tends to attract highly individualistic people who need a flexible organization to allow for their idiosyncrasies. They often struggle to maintain a sense of collectivism. Bahro claims that green parties often resemble 'a pandemonium of Thomas Mann's "lost citizens" in search of more sense of community than a political party will ever offer' (1986: 163). The emphasis on collectivism within green parties therefore sits uneasily with those aspects of green politics which emphasize the centrality of personal liberation and non-conformity. There also remains a paradoxical attitude to personal empowerment within the greens, because green parties 'believe in the power of individuals to change their situation, yet they are afraid of the power of individuality and of charismatic individuals' (Papadakis, 1984: 61).

Perhaps because of these paradoxes, and in spite of the ideals envisaged by the greens' alternative *modus operandi*, green politics tends to be conflict-ridden and rife with personality battles. Prominent party figures are consciously sidelined, which is one of the reasons the Greens in New Zealand have never countenanced the appointment of a single party leader. The rejection of 'personality politics' leads to an air of mistrust towards any party figures who might gain media attention or a public profile in the course of their political activities. Green co-leader of the Alliance, Jeanette Fitzsimons, recently stated that Green voters 'are disillusioned with the old, tired style of party politics . . . the strong leader who makes decisions for you and saves you having to think for yourself' (*Greenstone* December 1992: 21).

The confused attitude towards leadership is one of the reasons why lines of responsibility within green parties are unclear, allowing party processes to be manipulated by strong individuals and underhand tactics. Writing in a German context, but in a manner which reflects green party experience everywhere, Kolinsky (1989: 6–7) observes that: 'the internal party culture of the Greens is one of acrimony and conflict between competing camps. . . . The personalized nature of policies fuels personal rivalries and inner party conflicts. . . . [and] The members, who should have a substantial say if Basisdemokratie [basic democracy] were to work . . . seem to have been by-passed in a free-for-all of organizational uncertainties and personalized approaches'.

There is a significant discrepancy in most green parties between the benefits claimed as a result of their anarchic organizational methods and the actual outcomes of such organizations. Informal élites, the undue influence of particular individuals, a lack of clarity over responsibility, and the exclusion of members who do not have unlimited time to spend at meetings are some of the skewed outcomes which occur. The participatory democracy advocated by green parties at all levels is poorly promoted by the examples of green parties themselves. The Values Party, while putting considerable energy into the development of a party constitution which disempowered the party leadership, was still controlled by a small clique of members. These included paid party employees and a handful of committed activists who had the time and inclination to fathom and use the party processes. There is, in fact, little in green party practice to challenge Michel's iron law of oligarchy or Lawson's (1976: 227) analogous claim that 'All parties are run by the few'.

Conclusion

The essence of green politics is the search for a holistic paradigm to guide society's future development. This approach is not unproblematic, however, given the constraints of liberal democracy. For to impose a total package onto the political system is to threaten one of the basic safeguards of people in secular democracies: the right to private thoughts, opinions, beliefs, and values.

Holistic philosophy may be of relevance to particular individuals involved in, or supportive of, green politics, but the authoritarian implications of totalizing philosophies should not be overlooked. Both communism and fascism believed that they possessed the 'one true faith'; 'a comprehensive worldview, encompassing most, if not all, aspects of familial, social and political life' (Eisler, 1988: 181). Greens should not overlook the fact that, for those who have experienced communist or fascist regimes, 'terms like wholeness, totality and even community have perilous nuances' (Ryle, 1988: 11).

The solution to this problem is straightforward. Green politics, while motivated by a holistic paradigm, must focus its energies on the development of realistic policies which promote the immediate application of the concept of sustainability to all human activities. Green politics has an important and potentially pivotal role to play in formulating policies to address current dilemmas. Green politics has to demonstrate that it represents more than a vague set of values about how we should live with each other and with the natural world.

Green politics presents a comprehensive critique of the crises of modern society. It makes visible a range of concerns which are currently marginalized from decision-making processes. There is no green society, there is no logical conclusion to the green agenda. But there is a green perspective which can now be used in making all decisions about the future of society. The challenge is to ensure that the holistic green paradigm provides the basis for practical policies which make a concrete contribution to future development, rather than inspiring a vague and potentially totalitarian utopianism.

3   The Achievement of a Sustainable Society

> The history of popular movements . . . shows that only an arduous, even a tragic,
> understanding of life can justify the sacrifices imposed on those who challenge the
> status quo. (Lasch, 1991: 79)

If the green ideology could be captured in a succinct statement, it might best be
described as encapsulating the desire for a 'sustainable society'. Whereas in the
1970s greens embraced the concepts of 'no-growth' or 'zero-growth', in the
wake of the United Nations *Brundtland Report*, the concept of 'sustainable
development' now dominates the green agenda. This chapter argues that the
application of the concept of sustainable development will amount to a
fundamental reversal of current patterns of human activity in modern industrial
society. The achievement of sustainable development is a central task of green
politics, yet the method for accomplishing such a reversal of existing societal
direction has been inadequately canvassed.

Towards a Sustainable Society

Sustainable development addresses the need for current economic activities and
'development' to be compatible with the rights of future generations to non-
depleted resources and a healthy environment. It demands that social and
environmental issues, as well as the rights of future generations and developing
nations, are taken into account when decisions are made. Sustainable
development also explicitly addresses the need to bridge the gap between
'developed' and 'developing' nations. Specifically, sustainable development has
been defined as development which 'meets the needs of the present without
compromising the ability of future generations to meet their needs' (MacNeill,
1990: 113).

The concept of sustainability has been incorporated in planning and resource
legislation, such as the Resource Management Act in New Zealand. This Act
has, as its primary goal:

> the sustainable management of natural and physical resources [which is defined
> as] managing the use, development, and protection of natural and physical
> resources in a way, or at a rate, which enables people or communities to provide
> for their social, economic, and cultural wellbeing and for their health and safety
> while
>
> (a) Sustaining the potential of natural and physical resources . . . to meet the
> reasonably foreseeable needs of future generations

(b) Safeguarding the life-supporting capacity of air, water, soil, and ecosystems

(c) Avoiding, remedying, or mitigating any adverse effects of activities on the environment. (Resource Management Act 1991: 21)

The Act empowers local government, in particular, to take a strong line on the establishment and enforcement of strict environmental standards to alleviate the adverse effects of any development. But a problem is the lack of understanding by many local authorities of the concept of sustainability, which is one of the reasons why a concerted green presence is essential at this level of government. Unless those committed to achieving a sustainable society actively use the empowering provisions of the new Act, there is a danger that business will continue as usual, even though for the first time anywhere in the world this legislation requires that economic growth must be environmentally sustainable (James, 1992: 102).

Sustainable management implies the responsible 'management' of resource use. Sustainable 'development' extends beyond this relatively narrow concept to include, for example, issues surrounding the sustainable management of capital, as well as issues of global wealth distribution. But while sustainable development is a key platform of green politics, the implications of the application of the concept have not been studied. Even the *Brundtland Report* failed to examine the practical implications of sustainable development. Taken to its logical conclusion, sustainable development represents a fundamental challenge to existing cultural, political, and economic arrangements. Particularly in a New Zealand context, where a form of development economics has dominated public policy for much of our Pākehā history (James, 1992), the idea that development has to be sustainable challenges prevailing economic expectations.

The challenge posed by sustainable development has been succinctly captured by Alan Miller (1991: 79), who asserts that: '. . . global resources cannot be subject to sustainable development policies without massive change in the political and economic structures of the world'. Ruckelshaus (1990: 126) argues that moving towards sustainable development 'would be a modification of society comparable in scale only to two other changes: the agricultural revolution and the industrial revolution'.

Environmentalism has already challenged established patterns of development, making them a choice, and therefore 'something subject to revision' (Paehlke and Torgerson, 1990: 285). Radical environmentalists, as discussed above, suggest that the symptoms of the current eco-crisis point to the need for a radical transformation of existing society and its power relations. Paul and Anne Ehrlich (1990: 291) argue that the problem of species extinction, for example, 'will not be solved by minor adjustments to the sociopolitical system'. Commoner (1990: 117) is more explicit, asserting that if we wish to solve the environmental crisis the ideological issue of the extent of private control of natural and economic resources cannot be evaded. He cites the case of the 'Valdez Principles', introduced by a coalition of environmental groups after the Exxon Valdez disaster in Alaska. The Principles encouraged the belief that business has a direct responsibility 'to seek profits only in a manner that leaves the earth healthy and safe'. Such claims, Commoner notes (1990:163), 'fly in the face of [American] capitalism' and would require the corporate executives who possess enormous personal wealth and political power 'to relinquish a good deal of both'.

But if environmentalism alone challenges traditional development patterns, the concept of sustainable development provides an even greater challenge, demanding that both inter-generational and international distributive rights are addressed. The idea that extraordinarily powerful international corporations will willingly accept the diminution of their powers and reduction in their profits implicit in environmentalism and sustainable development is simply naïve. In fact, as Schrecker (1990) illustrates in a work of rare insight into the problems of environmental reform, contemporary business resistance to environmental reform is analogous to their earlier opposition to labour reforms. Business has actively campaigned against the 'unproductive' expenditure of capital on measures like pollution control demanded by new environmental requirements. Pollution control and other environmental safeguards are often still seen as impediments and obstacles to the growth which drives the industrial system. This is not to deny the efforts which many enlightened businesses are now making to ensure that their activities are more sustainable. But the green project for a sustainable society will not be achieved without opposition, for it challenges many vested interests.

The commitment to growth and expansion of the industrial system clearly conflicts with the notion of 'sustainable development'. Current patterns of development are *unsustainable* if they continue at current rates of resource depletion, pollution, energy use, etc. The industrial period of history has represented a shift from the use of current surface sources of solar energy, to the use of stored subsurface forms of energy, such as oil and coal. We can choose the rate at which we deplete the stored sources of energy, but the market has not adequately addressed this issue because it has no mechanisms for taking the future into account.

Many examples from the fishing and forestry industries in New Zealand, where resources have been greedily exploited for short-term gain with unfortunate long-term consequences, illustrate this point. In fact, the whole economic history of New Zealand, at least since European settlement, has been based on the unsustainable exploitation of natural resources, through fishing, (sealing and whaling initially), logging, mining, and pastoral farming. New Zealand's extractive economy has been a model of unsustainable development, reflected by the fact that before European settlement forest covered approximately 70 per cent of New Zealand—now it covers only 22 per cent:

Felling the native forests in New Zealand has been called one of the most ruthless and rapid changes which man has produced in the vegetation of the earth. (*Environment Science Technology* 1983: 522–8)

The parallels with the massive destruction of rainforests now occuring in places like Brazil and Malaysia are clear. But, whereas the current destruction of forests elsewhere is happening at a time of growing international awareness about the negative effects of such radical transformations of the natural environment, there were no such checks on the massive destruction which took place in New Zealand, at least until the 1970s when the Native Forest Action Council was formed.

In other areas, New Zealand's environmental performance continues to be abysmal, as the director of the Royal Forest and Bird Society, Kevin Smith, notes:

We continue to exploit our deep water fisheries to the point of extinction and to slaughter marine mammals and seabirds. We are still one of the most energy-inefficient countries in the world; we still rely heavily on fertilisers and pesticides to keep our agricultural economy going; we still bring new pests into New Zealand knowingly or accidentally all the time. (Smith in McVarish, 1992: 13)

New Zealand is home to 11 per cent of the world's endangered bird species, and a lack of adequate resource inventories makes an accurate appraisal of the total number of endangered plant and insect species difficult to establish. Few major waterways remain which have not been harnessed for their hydroelectric potential, while water pollution from agricultural chemicals is a major problem and many cities (including the capital, Wellington) continue to discharge raw sewage into the ocean. New Zealand's per capita waste disposal and energy use figures are among the highest in the world, and erosion—the result of deforestation and over-grazing—is a major problem.

Since the 1960s, however, there has at least been consistent environmental input into policy-making processes, both from a growing range of non-governmental organizations (NGOs) and from governmental organizations such as the Commission for the Environment and, more latterly, the Ministry for the Environment. Public concern at the ongoing loss of native forests was one of the main inspirations for the environmental movement which arose in New Zealand from the late 1960s. The continued destruction of New Zealand's native forests was also the subject of the second biggest petition ever presented to the New Zealand Parliament: the Maruia Declaration, containing more than 341,000 signatures and declaring a concern for the future of New Zealand's indigenous forests, was presented to Parliament in 1977. Issues regarding the protection of remaining forests have recurred regularly since, particularly in areas like the West Coast of the South Island.

Whereas there are unparalleled efforts on behalf of environmental protection, there is still a pervasive consensus about the benefits of industrialism and growth in modern, industrial societies like New Zealand. In recent decades one of the most graphic examples of the adverse effects of an uncritical attitude towards industrialism occurred in Japan where it took several years for the mercury pollution of Minamata Bay in Kyushu to cease, even after the source of the pollution was established. People's health was seriously and permanently affected after eating poisoned fish, but even the offending company's labour union supported the company's gradual response to ceasing pollution rather than an immediate stop to the harmful activities. The interests of pollution victims were considered subordinate to the basic aim of maintaining jobs in the district (Stockwin, 1982: 240).

Such disregard for human health and the environment can only be explained as the results of a system in which pollution is seen as a legitimate trade-off for the benefits of industrialism. The maximization of inputs (raw materials) and outputs (e.g, pollution) has been the hallmark of industrial era economics. Greens argue, by contrast, for the need to minimize throughputs. Such an approach becomes widely supported throughout the community at times of temporary crisis, such as the oil crisis of the mid-1970s, or the recent electricity shortages in New Zealand. However, it does not take long for the economy to return to its usual profligacy once the short-term shortages are overcome. A return to 'normal' economic activity is justified on numerous grounds, including the need for jobs and the generation of sufficient economic activity to ensure funding for the maintenance of a welfare state.

Sustainable development demands that there be criteria for determining the worth of production, taking into account the extent of the throughputs and the usefulness of the goods which result from the production. Encouragement should be given to those forms of production which, as well as minimizing throughputs, enhance the quality of life, human health, and the environment. Now that industrial society has reached a degree of consumer saturation in basic goods, we need to start paying greater attention to the quality of goods, rather than simply to the quantity of them. Products which meet social needs for health and well-being should be encouraged. Giving a new emphasis to aesthetics will encourage the production of goods which are beautiful or which make our homes or communities more attractive.

The paucity of alternative perspectives on the current malaise speaks volumes of the need for new and challenging approaches to economics and development. The current economic crisis in most modern industrial nations like New Zealand is seen as a crisis of consumer confidence. Politicians and economic commentators look anxiously for signs of improvement in consumer confidence indexes. People are not spending enough, and the solution is simply to get them spending again so that they buy the goods in the shops which means the factories will begin employing people again and the economy will resume its normal pattern of growth. But is this solution sustainable? I would suggest not. A commitment simply to encourage growth, without any criteria attached to what form of consumption will result, is simply going to continue unsustainable economic growth with all the associated adverse environmental and social consequences. Yet both major parties in New Zealand compete over claims to be pro-growth, the opposition Labour Party having recently announced that its priority as a government would be 'growth and jobs' (*Evening Post*, 29 February 1992). The Earth Summit of 1992—with its Agenda 21 for sustainable development—appears to have done little to elevate the banal level of political debate in New Zealand. Yet the current situation confronts greens with the prospect that a low-growth economy allegedly facing a 'recession', may be an example of a 'sustainable economy'.

Economic Growth

Economic growth is still commonly held to be the best guarantor of a better society. But a blind faith in conventional forms of growth is inconsistent with sustainable development as long as growth is based on the exploitation of natural resources and the increased consumption of non-renewable sources. Quantitative growth, when based on natural resources, simply cannot continue ad infinitum. Some economists are now claiming that we are entering a period where 'growth beyond the present scale is overwhelmingly likely to increase costs more rapidly than it increases benefits, thus ushering in a new era of "uneconomic growth" that impoverishes rather than enriches' (Daly and Cobb, 1989: 2).

Yet the global economic system is locked into an unsustainable growth treadmill, with anything less than steady economic growth interpreted as an economic downturn or recession, the only solution to which is a renewed period of growth. Economic growth becomes an end in itself as nations, regions, cities, and towns compete with each other to attract the transient currency of transnationals looking for new places to invest. As a former mayor of Detroit said: 'This suicidal competition among the states has got to stop but until it does, I mean to compete. It's too bad we have a system where dog eats dog and the devil takes the hindmost. But I'm tired of taking the hindmost' (in Logan and Molotch, 1987: 290).

The victories of some localities or nations limit the possibilities of others in a global competition which defies rationality and appears out of control. A form of Dutch auction operates as competing localities offer the lowest wages, the least restrictions, the greatest opportunity to despoil the environment without penalty because 'local authorities and national governments have accepted or been forced to accept the logic of competition for transnational investment' (Mackintosh and Wainwright, 1987: 259). In a period of economic downturn, as currently exists, regional rivalry (nationally and internationally) will grow, as not only business, but also the State, faces financial crisis.

In Australasia, the casino has become the symbol of growth with all its associated glamour. Such ventures are usually funded by foreign investment, bringing with them the promise of jobs and economic activity in the cities which welcome them. The criteria used to judge new investments such as casinos are often grossly inadequate in terms of appraising the social and environmental consequences of particular projects. More important considerations may include the ability of local politicians to take credit for attracting investment and jobs to their locality, claims which are enhanced by the construction of highly visible projects such as casinos. Because their political futures depend on it, local politicians become captives of capital. Controversial capital-intensive projects have been given city council approval in Wellington because councillors were left under no illusions that the investors were prepared to take their money elsewhere if the Council failed to approve their scheme. Any sense of ethical criteria, or future planning for a particular vision of society, is lost in the local politicians' attempt to attract a major investment to their locality. Any sense of judgment about what is worthwhile to produce or the process of manufacture and its impact on resource utilization or the workforce is abandoned in the rush to facilitate 'growth'. In this way, decisions made at the local level are crucial to facilitating the unsustainable growth upon which modern economies are based, which is another reason for a concerted green presence on local authorities.

Growth is not, of itself, undesirable and can under certain conditions be sustainable. It is the nature and quality of the growth which are important. Increased activity based on the exchange of information or enhanced social services need not have adverse environmental impacts and can be socially desirable. The growth of technologies with the potential to serve social or environmental ends, or the growth of services which enhance personal fulfilment, are both desirable and *sustainable*. Video technology, for example, has enabled the plight of the Indians in the Amazon rainforest to be broadcast around the world to an unprecedented degree. Similarly, technological advances meant that the massacre of students in Tiananmen Square was broadcast around the world almost immediately. Had such technology been available fifty years ago, might the Holocaust have been prevented?

Groups like the Alternative Technology and Lifestyle Association (ATLA) in Wellington are actively involved in bringing together the necessary actors, including the business community, to realize the potential of environment-friendly technologies. The Wellington City Council's Housing Division has started building energy-efficient homes and implemented energy-saving measures in existing housing stock. Courses about ecological housing, organic agriculture, and renewable energy options are now being taught in New Zealand tertiary institutions. All of these activities show that the green agenda for sustainability is not an abstract concept but is already being manifested in many ways throughout the community. The challenge is to make these examples of sustainable economic activity the basis of all levels of economic activity from now on.

Alternative Patterns of Development

Greens must begin to articulate and demonstrate the feasibility of 'alternative patterns of development' (Paehlke and Torgerson, 1990: 288). The many small-scale solutions, such as electric cars, sustainable energy-generating technologies, and natural health techniques, need to be brought together in a coherent package to suggest a way forward for modern industrial societies. Together, these activities demonstrate a 'green praxis'; tangible examples of how a greener society might look.

Greens should devote more attention to publicizing (and providing resources for) existing successful examples of alternative ventures, particularly in the business sector. The Body Shop, for example, has never spent a dollar on advertising and yet has been one of the most successful companies of recent years. The business's founder, Anita Roddick, asserts (1991) that: 'The role of business is not to create profits but to create live, vibrant, honourable organisations with a real commitment to the community'. She challenges modern business ethics and draws inspiration from those whom Campbell Bradley (1987) describes as 'the enlightened entrepreneurs'. This includes businesses such as Levers, Cadbury's, Boot, Huntley and Palmers whose founders, often out of religious conviction, created ethical businesses based on honest products, improved conditions for workers, and a philanthropic outlook. This business creed, as Roddick notes, 'seems long forgotten', although it does exist in some cooperative enterprises and alternative ventures. Business can be successful while taking an enlightened approach to issues which do not necessarily appear on the accountant's ledger. A rediscovery of the rich and successful history of ethical business provides useful signposts for the transition to sustainable development. The provision of seeding finance for ventures of this kind should be another ingredient in any package designed to facilitate the transition to a sustainable economy.

There is no shortage of things to be done, of worthwhile human activities to form the basis of sustainable development. Different social values based around conservation, reuse, and self-development will, of themselves, generate many worthwhile activities. These may not all be as visible as the kinds of 'megaprojects' which have characterized industrial development to date (e.g., roads and dams undertaken by the former Ministry of Works). The fascination with visibility is itself a product of the industrial culture with its concentration on the external works of humankind (Schneiderman, 1988: 26). The dam, with all its associated earthworks and construction, has epitomized the infatuation with large-scale, highly visible, environment-altering projects in New Zealand. A sustainable future may see such activities replaced by those based upon preserving and enhancing that which we already have: the natural environment, people, and our heritage. Roading engineers who have devoted their energies to providing roading solutions to traffic problems will have to turn their talents to constructing light rail systems and improved cycling facilities. Construction engineers who have specialized in constructing new high-rise buildings of dubious aesthetic value may have to learn building conservation skills in order to conserve and reuse existing buildings. As Raisa Gorbachev said on visiting a Swiss clock museum: 'This is what we should be doing, restoring things, not destroying things' (Sheehy, 1990: 195).

In Wellington, for example, the old BNZ Buildings are of heritage significance beyond New Zealand's shores. Yet they are dilapidated and at risk of being demolished. Their preservation and reuse could be providing many jobs, for skilled and unskilled alike, at the same time as meeting heritage conservation objectives and rejuvenating the inner city. I can think of few better examples of the potential for sustainable development, particularly if the full costs of options such as demolition are taken into account. Costs usually exclude the emissions from demolition equipment and truck journeys to landfills, not to mention the filling up of landfills (which is usually detrimental to the surrounding environment) with materials (wood, bricks, and concrete) which are recyclable. If the market made the true costs of many of its activities transparent, different decisions would frequently be made about particular projects.

Most of the technological solutions to current environmental problems already exist. What are lacking are both the financial incentives and the legislative requirements to introduce such technologies. Nor will they exist until the State takes an active role in establishing time frames and incentives for industries to adapt to the era of sustainable development. A primary role of the green movement must be to push for the kind of legislation which makes the new environment-friendly technologies viable and which ensures that the costs of introducing environment-friendly technologies are therefore shared across industries rather than by individual producers who show a willingness to respond to the demands of the environmental crisis. German legislation requiring cradle-to-grave responsibility by manufacturers for their products now means, for example, that the majority of components used in the manufacture of any Volkswagen car are recyclable at the end of the vehicle's life.

The role of legislation is very important. Strict emission standards for motor vehicles, for example, will have the effect of making it viable to mass-produce electric cars. Without the necessary legislation, however, there may be no incentive for manufacturers to overcome the costs of producing new technologies. Because the development and manufacture of environment-friendly technologies can still be counted as a cost in conventional economic terms, there is a need for the burden of creating appropriate new products and technologies to be shared more equitably if it is to become a social objective. This is what legislating for certain environmental standards achieves in practice.

Legislation is important to achieve a sustainable economy, and need not detract from a market-based economy. There have been few successful alternatives to the market-led economy. Nevertheless the market is not, as its advocates claim, simply the collective expression of individuals' consumer choices. People's choices are manipulated by producers in order to sell their goods. Nor is the market as democratic as its proponents claim, for the greater the capital, the more control one has in the market-place. But alternatives to the free market, like the late Soviet system, have proved grossly inefficient and unable to meet the most basic of people's needs. This is not to ignore the vast waste and inefficiencies in market-based systems, but simply to suggest that if alternatives to the market are to be discussed, their chequered history to date needs to be openly acknowledged.

It is important, therefore, in establishing the appropriate legislative and fiscal environment for the achievement of a sustainable economy, that there be a high degree of 'transparency' in what greens are trying to achieve. For example, if a tax is applied to all carbon-based products (a carbon tax), it needs to be explicitly stated why this is occurring, what the intended outcomes are, and what performance measures will be employed to gauge the success of the tax. If the tax does not work or produces skewed effects, there needs to be sufficient flexibility for the policy to be changed. If it is deemed that the production of wheelchairs is more important than the production of armaments, then the criteria for having made this decision need to be explicit and clearly accessible, with financial repercussions presented transparently. This involves a sophisticated process of prioritizing green goals, clarifying objectives, and acknowledging different ways in which they might be achieved. Many modern business management and other techniques (including 'envisioning' and other New Age skills) provide appropriate frameworks for the kind of goal clarification process which green politics needs to embark upon.

An active and aware consumer movement will be a vital component of green attempts to promote alternative patterns of development. Publications such as *The New Zealand Green Guide* (Davis and Hodge, 1990) provide consumers with access to a wide variety of environmentally friendly options, in areas ranging from consumer goods to gardening and health care. One movement in Wellington which has shown the power of 'green consumerism' is the Residents' Airport Noise Action Group. This group has encouraged people to travel on the airline which creates least noise pollution for the neighbourhoods adjoining the airport. Ansett had deliberately bought a fleet of 'whisper jets' because of the issue of airport noise in Wellington, and the pressure exerted by the residents' group has recently seen Air New Zealand respond (after several years) by beginning to fit hush-kits to its own fleet. Without active public campaigning on this issue, it is unlikely that Air New Zealand would have acted to alleviate the noise pollution.

Further Green Perspectives on the New Zealand Economy

New Zealand's current pattern of development simply mimics the model of unsustainable development practised in most 'developed' nations. No government in recent years seems to have put the protection of New Zealand's unique quality of life and the environment at the centre of the political decision-making process. There is a lack of vision about what is special about New Zealand, which is one reason we copy the development patterns of modern, industrial nations elsewhere rather than defining our own role in the world. And yet, identifying what is unique about New Zealand and, in particular, what is worth preserving, might help us to chart our future direction as a society more clearly:

> In international terms, there's a stronger commitment and more opportunity for looking after the environment here than anywhere else in the world. As David Bellamy said: "If we can't get it right here, then nobody will". He's right. We start so much further ahead than most places: small population, high standard of living, and we're approaching energy self-sufficiency. (Johnston, in McVarish, 1992: 20)

New Zealand could provide a model to the world of a sustainable economy, particularly in the fields of agriculture and forestry where we have significant experience. Such a 'vision' would capitalize on existing areas of strength and provide a competitive advantage which would have positive spin-offs in areas like tourism. However, New Zealand appears unable to mobilize behind any particular vision of what kind of a society it wants and where it is going. As well as reflecting the short-term focus of New Zealand politics, this reflects several other factors, including our relative youth as a post-Treaty of Waitangi nation and, relatedly, the confusion about Pākehā identity. Perhaps when Pākehā are able to gain a clear idea of what it means to be a Pākehā, we will be able to contribute more clearly to a vision for New Zealand.

The creation of a unique vision for New Zealand will be assisted if we consider why it is that there are large numbers of people in the northern hemisphere who want to emigrate to New Zealand. What attracts these people to New Zealand is its isolation and relatively unspoiled environment. It is these unique attributes which we should seek to preserve and enhance, making them a focus of our current and future policies and decisions. New Zealand is doing poorly competing on the terms of the big nations, so we should build upon our strengths as a clean, green country which has gained an international profile in promoting anti-nuclear policies. Such a vision can provide the basis for a sustainable economy.

The defence of New Zealand's relative uniqueness is a potentially powerful force which has inspired some of the major environmental issues in this country. Foremost among these battles was the campaign to save Lake Manapouri—'a symbol of nature despoiled in the name of progress' (Cleveland, 1972: 96)—from hydroelectric development. It is also worth noting that many people in the Values Party during the 1970s and in the contemporary green movement have either travelled and lived abroad, or else consciously made New Zealand their home after migrating from the industrialized northern hemisphere. They are people 'who seemed intent on trying, after getting away from it all, to stop it coming here' (*Critic*, 6 May 1975).

The number of 'environmental refugees' who have moved to New Zealand (or who still aspire to) should enable us to see our peripheral status more positively, as an environmental haven in the South Pacific. A vision of New Zealand as a laboratory of sustainable development for the world would build on this nation's historical record as an innovator in the field of social policy—and more latterly, as a nation not unafraid to take a stand against nuclear proliferation—which would ensure us a unique niche in a highly competitive world.

By contrast, the re-establishment of a 'fortress New Zealand', as many of those in the green movement and in the Alliance desire, will not produce the kind of outcomes I believe the green movement should be seeking. On the question of economic insulation through import controls, for example, it is worth citing the case of workers in the London furniture industry who were brought together by the Greater London Council to discuss the sad state of their industry. Most participants initially had a favourable attitude towards import controls and the subsidization of local industry. However, their favourable attitude soon changed as they realized that, without addressing larger issues, such as the control and direction of the industry, import controls and subsidies 'would merely provide a support for the present management with all its incompetence and lack of care and imagination' (Mackintosh and Wainwright, 1987: 231). Furthermore, as a long-term solution, 'import controls would provoke retaliation and an even worse financial crisis: working-class people in the West would suffer but the economic misery in countries like Poland and Ethiopia would be even worse' (Mackintosh and Wainwright, 1987: 231–2). London furniture workers realized that there are broader issues than saving existing jobs in declining local industries, that protected economies—like New Zealand's prior to 1984—create privileged élites at home and ignore issues of international distributive justice. Improved sectoral marketing and cooperation (as in the Swedish furniture industry), along with building on existing competitive advantages, are some of the alternatives to the tendency towards insulation.

Inter-sectoral cooperation and forward planning involve a degree of future planning which has been actively discouraged in recent years. New Zealand has abolished all the institutional means by which it could ensure the provision of longer-term planning essential for sustainable development. Since the abolition of the Commission for the Future and, more latterly, the Planning Council, there have been no statutory bodies charged with taking a longer-term perspective on the likely domestic and international changes which will effect this country. Yet there is ample evidence of where long-term planning could have been beneficial, whether in terms of energy-demand projections, population policy, or responding to British membership of the European

Community in 1973. Clearly, there is a need to put in place procedures able to predict with a degree of certainty the impact of events such as the depletion of the Maui gas reserves. A system of forward planning across all activities of government should be established to counter the inability of the market to take the future into account. This system should also be charged with responsibility for maintaining a national resource register and positing different future scenarios, making it clear that there are choices in terms of the direction in which society develops. It is simply not true that 'there are no alternatives', as conventional politicians have argued in recent years.

The Difficulty of Achieving Change

I have attempted thus far to discuss what sustainable development is and to identify some specific examples of the kinds of activities to which it might give rise. But the difficulty of achieving a new (sustainable) pattern of development needs to be addressed explicitly. Difficulties in achieving change include identifying what groups and movements, if any, are currently equipped to put into practice such a massive reversal of current economic and social arrangements. Traditional change theories have looked to particular historical agents (e.g, the working class) to achieve their desired changes. But who will drive the movement towards a sustainable society in the contemporary world?

Social democracy and labour politics have been the traditional focus of radicalism in this century, but they appear unlikely harbingers of new patterns of development. Socialist politics and the labour movement have traditionally tempered the excesses of the market economy in the pursuit of particular social objectives. However, the experience of social democratic and labour governments around the world during the 1980s, including the Fourth Labour Government in New Zealand, suggests that they no longer offer solutions to the destructive dynamic of the current industrial system. Indeed labour politics has clearly become wedded to the current system of industrial growth as the best means of achieving the improvement in the conditions of its traditional constituency. Labour politics has become wholly assimilated into modern industrial society:

> The labour movement is too imbricated in the structures, institutions and dilemmas of the past, defending old interests and social relations, standing in the way of a 'new project'. It is fettered by its commitment to economic growth, to the state (in its Keynesian welfare state variety), to the reality, ideals and norms of industrialism. (Olofsson, 1988: 16)

To the extent that labour politics still seeks reform, it looks to sponsorship from some of the worst practitioners of unsustainable development to fund its programmes, as Labour governments on both sides of the Tasman demonstrated during the 1980s.

Furthermore, the demands of the traditional labour constituency are now essential to fuelling the industrial system 'through their appetites and desires':

> No wonder the antagonism they once exhibited towards the wealthy has mitigated; they are united with them to fuel the demand for continuous growth and expansion, seeing their best hope in the ever-increasing purchasing power that threatens to eat up the earth and everything on it. (Seabrook, 1990: 167–8)

Little wonder then that social democracy should have so little to say on the dynamic of the industrial system which is destroying the very natural world on which all economic systems ultimately depend: '. . . the most bitter of silences that have descended where debate and argument should now be more passionately engaged than ever' (Seabrook, 1990: 168).

Nevertheless, the labour movement did pose a substantial challenge to unchecked economic activity for a significant part of this century. It is difficult to identify a movement today which poses the same threat to unchecked economic exploitation as the labour movement did. Writers such as Gorz (1982) argue that all the disenfranchised and marginalized groups will join together to form a powerful movement for change. But this simply represents a reformulation of idealistic Marxist schemes; it is a world devoid of individuals, but composed of 'forces', 'classes', and 'movements' based on a 'pseudo-scientific law' (Scruton, 1985: 68). In response to the suggestion that the marginalized classes may, anyway, simply want to mimic the aspirations of the privileged classes, Dobson (1990: 168) suggests that 'the Green movement could make the marginalized class aware that this habit has no future, and that its interests lie in a different form of society rather than immersion in the present one'. How, precisely, this disparate group of people are to be made 'aware' is not spelt out, as in most grand schemes which treat people as a mass.

It is difficult even to achieve a minor reform, let alone a major social transformation. The Minister for Transport, Rob Storey, recently faced concerted and widespread opposition to the imposition of a relatively minor tax on petrol to fund public transport. This example of an attempt to implement a minor part of the green agenda should alert greens to the opposition which will greet many of their policies. Rob Storey observed recently (in Campbell, 1992: 17), while New Zealanders may have a lot of sympathy for improved environmental policies in general, 'we're not doing very well on the things that involve uncomfortable personal choice', such as riding public transport to work instead of taking the car.

But perhaps the major impediment to achieving change in advanced industrial democracies is the degree to which governments are dependent upon the flow of private investment. Private capital sustains the economic growth which provides both income and employment for their citizens, as well as enabling the provision of State services of various kinds:

Thus, public policy in various areas must operate in a context of more or less continuous latent tension between the state's various other mandates, such as eliminating the various damaging effects of economic activity on the environment, and the need to sustain the conditions for capital accumulation— that is, to maintain a favourable business climate'. (Schrecker, 1990: 170)

The most powerful influence that business has over any administration which would temper its damaging activities is the threat of locating elsewhere; of removing jobs and sources of public revenue to another location which is less stringent in its regulations. The threat of a 'capital strike' or 'capital flight' is a very real one for many modern, industrial nations when developing low-wage nations such as Malaysia and Indonesia present such an attractive proposition for capital. Especially where a single industry dominates a town or region, this gives that industry tremendous power and influence to ensure that environmental or other regulations which might reduce its economic viability are circumscribed or not introduced at all. In this situation, it is difficult to make companies pay the true cost of their social and environmental damage. But, as Claire Nader argues, the unresolved environmental problem 'is a form of deliberate regulation' no less coercive, and perhaps more so, than the actions of governments which attempt to protect citizens from the consequences of environmental damage (in Schrecker, 1990: 186).

Environmental regulation such as New Zealand's Resource Management Act 1991 effectively seeks to reduce the prerogatives of private capital to exploit or despoil what are frequently collective resources such as air and water. By its very nature, environmental regulation challenges the power of business because it recognizes that resource allocation decisions cannot be left entirely up to the market, and that the community should have some say over the allocation of resources. Business has already shown that it is prepared to organize against even relatively minor reforms with regard to reducing CO_2 emissions and phasing out the use of ozone-damaging chlorofluorocarbons. Continued American resistance to making a commitment to reducing harmful emissions of agricultural and industrial gases, while cloaked in terms of the scientific uncertainty surrounding global warming, is really because of the potential costs to the American economy of the reduced usage of oil, coal, gasoline, and other major sources of greenhouse gases. When a power which contributes so much of the world's greenhouse gases has consistently baulked at making hard decisions because of domestic political considerations, the difficulties of achieving change should be clear for all to see.

Realizing how difficult it is to achieve even minor reforms, greens need to devote far more attention to issues of tactics and strategies. Unfortunately green politics, to date, has shown precious little concern with these prerequisites of effective political action. Observers of the greens in parliamentary settings have noted the extent to which greens have put energy into their rhetoric rather than into concrete policy initiatives. After seven years in the Bundestag, the German Greens achieved few legislative changes, although their impact cannot simply be measured in these terms. During their term in the Riksdag, the Swedish Green MPs, for example, were noted for asking the largest number of parliamentary questions and making the most interventions (Gaiter, 1991: 35).

The consciousness-raising role of green politics is important in its own right, and can lead established parties to take legislative initiatives. While concrete evidence is difficult to produce, the Values Party clearly contributed to a popular anti-nuclear sentiment in New Zealand which was eventually enacted in nuclear-free legislation by the Fourth Labour Government.

Political commitment will be central to creating a policy-making framework which addresses the conflict between sustainable development and the short-term profit-oriented goals upon which most modern business is built. In the pursuit of sustainable development Ruckelshaus (1990: 129) asserts optimistically that 'the values are there; the appropriate motivations and institutions are patently inadequate or nonexistent'. Modernizing our political institutions becomes a key issue.

The Challenge to Western Cultural Assumptions

Modernity is a distinctly Western project according to Giddens (1990: 174–5). Yet the green political project amounts to a fundamental critique of the established social, economic, and political arrangements in modern Western societies. The changes which green politics seek amount to a direct challenge to the notion of 'progress' which has dominated modern Western culture. The Western view of progress sees as inevitable the movement of society in a linear direction culminating in a perfect society. The belief in 'progress', which has been a defining characteristic of Western culture, sits uneasily with the concept of sustainable development.

Reading the literature which discusses a sustainable society, it is possible to glean just how great the challenge a sustainable society will pose to the existing culture is. Keekok Lee (1989), for example, in *Social Philosophy and Ecological Scarcity*, asserts that a lifestyle of frugal consumption is morally commendable. She calls for economic activity to be based around horticulture rather than manufacture, and proclaims the need for 'a conception of human needs based on the notion of sufficiency, not infinite ever-expanding insatiety . . . [and] an alternative source of abiding satisfaction which lies in self development' (Lee, 1989: 385). Human relationships, conversation, games, dancing, singing, 'internal goods' (self-development, for example), and active participation rather than passive consumption are the possible sources of satisfaction in a social order based on her interpretation of the principles of sustainable development:

> . . . we should seek satisfaction and worth in activities which consume as little of these scarce goods as possible, that we should, on the whole, prefer those activities which make less entropic demands rather than those which make more. (Lee, 1989: 209)

The model of society Lee envisages as best able to fulfil this vision is that of an ascetic form of socialism.

There are several recurring themes in the scenarios of social orders that are run according to the principles of sustainable development, of which Lee's is but one example. These include the desire for community, the moral imperative towards a low-consumption frugal lifestyle, an emphasis on self-discipline and restraint, a heightened sense of consciousness of others and the planet, a rejection of overt stimulation, the rejection of technology, and a high value attached to authentic relationships with others as well as with the natural world. This catalogue of anti-modernist sentiments amounts to a fundamental challenge to the way people live, and to many 'sacrifices' in terms of reduced material well-being and the subjugation of personal desires for the greater collective good. But history suggests that where personal sacrifices are made for the group there is usually a degree of exclusiveness which allows the group to see itself as distinctive or as united against a common enemy. Hence, there are the highly successful Mondragon community experiments in the Basque region in Spain (where Basque nationalism provides a strong sense of social cohesion), or the successful farmers' cooperatives in Finland at the time of Russian rule in the nineteenth century (when Finns were united against a common occupying force). But unless such exceptional circumstances exist, one wonders how the massive change in expectations envisaged by Lee and others will occur.

Lasch (1991), for example, has asserted that the values to which we need to return for a decent (and sustainable) society are recoverable from the 'anti-progress' lower-middle-class ethic of limits which played a large part in the past of colonial societies like New Zealand. In the 'impending age of limits', there will need to be a return to the belief 'that moral wisdom lay in the limitation rather than in the multiplication of needs and desires' (Lasch, 1991: 39, 45).

But how is the individual, accustomed to our high-stimulation, urban existence, to adapt to new pleasures based on personal growth and cultural and intellectual pursuits? Lasch's prescription is a collective one, based on the *hope* that people will be prepared to return to a lifestyle less ostentatious, but more spiritually fulfilling, than those they enjoy at present.

How, precisely, is this collective cultural shift to take place? Personal and cultural transformation cannot be imposed collectively. Modern attempts at creating perfection for an entire society have been a recipe for bloodshed, whether in the former Soviet Union or in Pol Pot's Kampuchea. The imposition of a singular set of norms and values onto heterogeneous populations must not be countenanced by anybody concerned about freedom or individuality. Spiritual decisions and options must be left firmly in the terrain of individual choice. What can be achieved collectively, however, is the provision of information so that all people are aware that choices do exist, that there are opportunities to grow, to change, to be different. People can also be presented with working examples of different ways of doing things and the choices to experience these. A cheap, efficient public transport system which got people to work faster than the individual motor car and which provided a pleasant experience might start to seriously compete with the private car, but we cannot know this until it has been tried. In the meantime, people cannot be forced to change, and political strategies which do not recognize this usually end in violence.

Changing Personal Values

Changing cultural norms and values amounts, in practice, to changing the behaviour of individuals. On a personal level, however, the transition to sustainable development will be no less traumatic than the changes at an institutional level. Yet proponents of change offer nothing more than the nebulous promise that in a sustainable society, '. . . the life it would offer would be strangely different from our own and humankind would become radically different' (Coombs, 1990: 59). People are exhorted to change, but with little indication of how this will occur. How, for example, are we to take up the frequent challenge 'to promote attitudes and behaviour which value simplicity and frugality and deprecate ostentation and waste' (Coombs, 1990: 13)? Coombs also suggests that we must abandon the notion that 'bigger is better', but the process by which society is to take this qualitatively different direction is neglected by his dissertation. Those who do address such issues usually revert to calls for a metaphysical reconstruction, or for a pervasive new consciousness based on religious values:

> It will be necessary for people to 'internalise'. . . [the] ecological doctrine so that it becomes part of the unconscious motivation which moulds their conscious thought and action. It is interesting to speculate on what manner of individual will achieve such a 'translation'. (Coombs, 1990: 51)

Bahro (1986: 104) is quite clear that the transformation will come from the transformed. Who will judge whether one is transformed or not Bahro does not discuss.

Change at the level of the individual alone will not be the basis of a cultural transformation, but it is nevertheless important to the green agenda. The effects of individuals growing in self-esteem, rejecting certain social norms in pursuit of their truly personal feelings and aspirations, and embarking on a spiritual journey should not be underestimated when it comes to formulating change strategies. Personal change can affect individual lifestyle choices, but the importance of change at the personal level for broader political and cultural change is also recognized in some more theoretical green writings. For example, deep ecologists argue that the great challenge is to broaden the boundaries of individual perceptions so that people might identify with the natural world and with other species. Deep ecologist Arne Naess's ecologically inspired normative system promotes self-realization because 'the higher the Self-realization attained by anyone, the broader and deeper the identification with others' (Naess in Fox, 1990: 103).

Self-realization corresponds to Rifkin's (1991) assertions of the need to 'remake ourselves', applying the lessons of psychology to society at large so that 'the new thinking about security flows from the inside out' (Rifkin in Polsgrove, 1991: 39). He suggests that the external search for material and strategic security is the cause of a plethora of modern ills, including militarism and the destruction of the environment. Analogously, Gloria Steinem (1992) argues for the power of self-esteem. While there is substantial grass-roots interest in self-esteem, she argues, it is suppressed by government, media, and religious establishments because 'the idea of inner-authority is upsetting to those accustomed to looking outside for orders . . . and certainly to those accustomed to giving them'. Self-esteem leads to empowerment and self-authority, a concept which Steinem asserts is 'the single most radical idea there is' (*Time*, 9 March 1992: 41). This idea sits easily within the green movement which, along with other dissent movements in modern society, voice '. . . above all, a plea to their fellow citizens to curb the impersonal powers that ruled their lives and thereby to reassert control over their own destinies' (Hughes, 1990: 149).

Without self-esteem, it is unlikely that people will be able to overcome the powerlessness and anomie so common in the modern world:

> There is a degree of harmony and wholeness to which men are entitled; yet modern society has consistently undermined every institution and social experience which could encourage such a flourishing of the individual. It has forced new conditions of life on its citizens—factory work, big cities, the cash nexus—which mutilate them in their innermost being. (Zweig, 1980: 246)

The resultant deadening of sensibility makes it possible for people to screen out the horror of nuclear holocaust or environmental catastrophe (Hughes, 1990: 4) and helps to explain why autonomy and what Phipps (1990) describes as 'our natural higher potential' and 'the full humanization of the person' have been consistently strong components of the green movement.

Giddens (1991: 208) argues that the emphasis on personal growth represents a major social transition in late modern society. Personal growth will be essential to harmonious interdependence on a global level. Giddens (1991) argues for emancipatory politics to develop a concept of 'life politics':

> Life politics concerns political issues which flow from processes of self-actualization in post-traditional contexts, where globalizing influences intrude deeply into the reflexive project of the self, and conversely, where processes of self-realisation influence global strategies. (Giddens, 1991: 214)

Personal growth is inextricably linked to solutions to the crisis of modernity. Personal growth, new lifestyles, 'and the requirements for a new civilisation are in harmony', according to Carol Riddell (1990: 27) of the Findhorn Community in Scotland: 'The process of reorientation towards inner awareness involves excitement, joy in living, growth in creativity, a relative release of material needs, increased ability to accept people as they are and a determination to resolve problems' (Riddell, 1990: 27).

Modern society encourages outer-directed (as distinct from 'inner-directed') behaviour, and busyness and activity instead of introspection and contemplation. The archetypal inner-directed person follows their own instincts and feelings, and consequently makes fewer demands on natural resources because they do not seek continuous stimulation through material consumption. Their lifestyle seeks to overcome the reduction in sensibility to nature and begins to reveal 'the characteristics necessary for the new human civilisation which will have to replace our present one' (Riddell, 1990: 34).

Is Change towards a Sustainable Society Possible?

Can a transformation which depends on a profound individual and cultural change actually be achieved? Is human nature really so malleable as those proclaiming a cultural—or religious—transformation would have us believe? Are the numerous new movements and activities taking place in advanced industrial societies now indicative of a generalized, qualitative change in values, or are the vast majority of people unaffected by the alleged manifestations of a new paradigm?

Perhaps the most common green prognosis of how change will occur is based on the optimistic assumption that a variety of forces—including changes in personal values—are at work which make the greening of society somehow inevitable. A local green activist recently wrote: 'We don't have to worry about our greenness because the greens' time has come, and it's going to have to be the other parties who have to turn green as time goes by . . . it's up to us stalwarts to keep the [Green] Party strong until slowly but surely the other parties realise that our ideas are right' (personal correspondence to the author, 9 June 1992). This optimism pervades green politics and is based on a deterministic view of change which believes in the inevitability of transformation in a particular direction. Given that green politics tends to the view of modern society as an aberration, the belief in the inevitability of a miraculous change of direction in social evolution may not be as paradoxical as it first appears.

However, the assumption that there will be an inevitable transformation of consciousness has been criticized from several quarters. Pepper argues that this view is driven by a neo-Malthusian assumption of environmental determinism and 'an almost Messianic belief in the historical *inevitability* of radical trans-formation to a "new culture" which is part of an evolutionary process as inescapable as Darwin's biological evolution' (Pepper, 1985: 15). Marien asserts that claims of inevitable cultural transformation are poorly supported, asking:

'Might it be possible that the Transformation is not happening at all, but that a fair number of middle-class professionals have been led into a religious trance with the soothing idea that it *is* happening?' (Marien, 1986: 69). He goes on to argue that 'the world crisis with a hundred names' still remains largely invisible to mainstream culture, or is readily dismissed as 'small is beautiful romanticism' (Marien, 1986: 58). Certainly the means by which people might gain access to the transformation may not be universally popular, or practicable. Bahro, for example, claims that 'clear-headed mysticism leads to a profound mobilisation of emancipatory forces in the human psyche . . . and should be made accessible to everyone, for example through the practise of meditation' (Bahro in Papadakis, 1984: 47).

Unrealistic assumptions about the ease with which new ways of thinking can be encouraged as a basis for political reform are not new in politics. The former President of the Soviet Union, Mikhail Gorbachev, began his reform processes of glasnost and perestroika with a concerted effort to change the way that the Soviet people thought. His efforts were unsuccessful, but in assuming that people were malleable and responsive to new ideas and suggestions he reflected a widespread belief in the green movement that sufficient extolling of the need for change will see people respond by willingly adopting new patterns of thought and behaviour. Such a perspective is naïve, to say the least, but also potentially very dangerous, for there will always be the problem of what to do with those who refuse to change their way of thinking in accordance with the ambitions of those pursuing change.

The belief in the inevitability of change also shows a lack of historical awareness. The Values Party, for example, saw its success growing exponentially from election to election until it would finally take its rightful place as government. Candidates consistently overrated their chances of winning votes and parliamentary seats, as well as overestimating the influence of their manifesto. One activist couple showed not untypical enthusiasm when they wrote to the party organizer about Values' 1978 'minifesto':

> The practical wisdom it contains alone should throw light into the eyes of the New Zealand masses and sweep the obscure but brilliant Values Party into power, so they may lead us into a new age of enlightenment through caring and sharing. (Values Party Archives, WTU 85/11, Box 10, Turnbull Library, NLNZ)

The ensuing 1978 election result saw the Values Party's vote decline to 2.8 per cent of the total vote, as compared to 5.2 per cent in 1975. Subsequent election results in Europe also demonstrate that there is no guarantee of a constant improvement in green political fortunes. Latterly both the Swedish and German Greens have lost their parliamentary representation.

An additional factor which requires caution in considering the potential for achieving the sustainable society is the mounting opposition to green ideas and to environmentalism. In Germany, a 'Car Party' was formed in direct opposition to the Greens. The development of other parties in a European context, most notably the anti-immigrant, extreme Right movements, demonstrates unequivocally that not all parts of society are moving in a green direction. Business is organizing against environmentalism in a coherent manner for the first time. In New Zealand there has recently been a concerted effort by certain sectors to discredit the green movement and green policies. The sentiments of the recently retired head of the Fishing Industry Association in New Zealand may be an indication of the hardened battles to come. Peter Talley (*Nelson Evening Mail*) claimed that: 'Environmentalists are stating a warped view of what society wants', and that they have a higher profile than their level of public support warrants. His calls for the abolition of the Department of Conservation parallel employers' calls for the abolition of the Labour Court. Economically difficult times have reinforced the growing opposition to the perceived threat which greens pose to development. The chief executive of the Contractor's Federation, Peter Tritt, recently wrote in a column full of vitriol (*Contractor*, November 1992) that green is 'not so much a colour, more a block to progress'.

There is no guarantee that society will continue to grow in the direction desired by greens. After the green movement faced several setbacks in recent Californian citizens' referendums, former State Legislator, Tom Hayden, proclaimed that 'California is now Paradise Lost. . . . That's very bad for the world. If a green revolution hasn't happened in California, it certainly won't happen anywhere else' (*Dominion Sunday Times*). Repeatedly we are confronted with the need for effective tactical and strategic thinking by the green movement. It is easier to repeat slogans about the inevitability of change than to actively work out goals and priorities and the strategies which might actually be successful in achieving them. This means an openness to all possible strategies which might pave the way to a green society, including the promotion of certain kinds of growth, anathema though that may be to many greens.

For example, if we take reduced population as a green goal, it must be asked why it is that affluent, urbanized societies have achieved a balanced population growth, the most advanced environmental legislation, and the greatest steps towards benign technologies. Hence Brundtland's (1990: 137) assertion that economic growth has the potential to provide the capital for solving environmental problems. It is growth, after all, which provides the wealth to address environmental problems, to raise educational levels—an important precursor to green support—and to address the redistribution of wealth both nationally and internationally. Anderson and Leal have commented that:

> Advocates of sustainable development argue that human betterment should be measured in terms of health, education, improved living standards for the most disadvantaged, and a cleaner environment. These conditions are precisely the results of economic growth. (Anderson and Leal, 1991: 171)

Social democracy has always depended on the success of capitalism for the resources to fund welfare and other social measures such as free and universal education. The green agenda is not dissimilar in that it will be more easily 'afforded' during periods of economic growth, and is likely to be funded by the profits of a prospering economy, at least for the foreseeable future. Thus we strike yet another paradox of green politics, but one which must be honestly addressed. In New Zealand this means addressing such issues as those concerning population for it can be argued that public transport and other desirable initiatives will never be viable so long as there is such a limited population.

Conclusion

Sustainable development is not an unproblematic concept, nor will its application be achieved without a great deal of resistance from vested interests. In a country like New Zealand, with an extractive economy based on unsustainable development, the challenge of achieving sustainable development will be considerable. However, this can be turned into a strength by making New Zealand an international laboratory for the development of environment-friendly policies and activities.

A sustainable society will be the result of a multi-level approach to change, ranging from conditions which will foster self-esteem to the enactment of environmental legislation and policies. At the level of the State, there is a need for policies and processes which place the attainment of a sustainable future at the political centre-stage. Government needs to put in place a legislative and financial framework which encourages sustainability in all sectors of society. It is hard to avoid the conclusion that a sustainable society will require an active role for the State, even if only in sending out the right messages to 'the market' through taxation and purchasing policies. Democratic input into investment decisions will also be necessary, given the propensity for capital to be wasted on speculative ventures such as the urban real estate boom in New Zealand in the 1980s, rather than upon investment in sustainable economic activity. Economic historian, John Gould (in James, 1992: 57) has written that 'New Zealand has something of a genius for wasting capital'. Sustainable development demands a rethinking of this propensity, for without such rethinking sustainable development will not be achieved.

Economic tools cannot address the individual dissatisfaction which drives unsustainable material consumption, however. But the personal growth movement can play a role here, and green politics should explicitly include the maximum distribution of self-esteem as one of its social policy goals. Nor should greens fail to notice the importance of the numerous spiritual pursuits which people are now embarked upon, or the conventional wisdom of the Māori, who decry of the Pākehā that: 'There are answers for our country within this country . . . and yet they will not listen' (Wihongi in McVarish, 1992: 19).

The achievement of a sustainable society must become the primary goal of the State's influence in every sphere of society. Architects now speak of green architecture, accountants of green accounting. Even the Automobile Association has recently released an environmental policy statement. There are examples of positive reforms, but also many superficial attempts at green window-dressing and much rhetoric unsupported by action. Even incremental reforms are difficult to implement, not just because of the complexity of the issues themselves, but also because environmental reforms pose a threat to the assumptions of infinite growth upon which modern business is built. It took more than a year, for example, to get a policy through the Wellington City Council banning the purchase of imported rain-forest timber by the Council. The policy was actively opposed by local embassies representing rain-forest timber exporting nations.

The lack of tactical clarity on the part of the green movement does not place them in a strong position to deal with obstacles to the achievement of a sustainable society. There is a need, therefore, for well thought-out green strategies and tactics, and a clear set of policy priorities and goals. An immediate goal must be for greens to take a proactive role in broadening the scope of what we call 'economics' so that the total costs of production are transparent, including those costs to the environment and the community. Means of encouraging socially and environmentally useful production must be promoted through the use of economic instruments. Only with such concrete policy mechanisms can the concept of sustainable development become more than a laudable goal.

If it is difficult to implement even relatively minor reforms in the direction of a green society, then the goals of those aiming for a mass cultural transformation or a total overthrow of the existing social order appear as the manifestation of wishful thinking. Feminist advances in the 1970s and 1980s are now being confronted with an anti-feminist backlash. There is no guarantee that green issues will not be susceptible to the same reversal in fortunes, and there are already indications of a concerted response from opponents to the green agenda. In the meantime realistic policies and strategies must be advanced which the majority of people can understand, and which are compatible with personal liberty and political freedom.

This must include publicizing changing working examples of a sustainable society, ranging from alternative health care, to organic farming, soft energy options, and environmentally friendly technologies, along with personal growth therapies and self-help groups. There is now in existence, in most advanced, industrial societies, a whole 'alternative sector', relatively unacknowledged, yet contributing beyond all doubt to the quality of life. One of the key roles of the green movement must be to give this alternative sector greater publicity and resources, so that it is visible to the general public as a model of how parts of a green future might look.

4   Concrete Tasks for a Green Future

> To be sure, choices in institutional design are not as grand and inspiring as the great ideological struggle between capitalism and socialism. But those choices are where the big public decisions of our time lie. (Starr, 1992)

This chapter will suggest specific parts of a green agenda for change which modern society must address. As has been said previously, change will occur on many different levels, in the daily life of the individual as much as in the conventional sphere of 'politics'. Perhaps *only* if it occurs on many levels will a sustainable society be achieved. Certainly political reform which has been focused primarily at the institutional level has not achieved the desired effect, while it is questionable how much impact those seeking change solely on the personal level—through the New Age movement, for example—have had on the broader political environment.

The Reform of Political Institutions

The very presence of green politics in modern, industrial societies means that basic questions about the kind of social organizations that we have and the values they are predicated upon are open to question. The evidence of the breakdown in the human–nature relationship alone, 'requires a major personal and collective re-evaluation of the meaning of human life on earth, and of the source of happiness' (Riddell, 1990: 15). This re-evaluation applies as much at the level of our political institutions as it does to the individuals in whose interests these political institutions are supposed to govern.

Much of modern politics, however, can best be described as technocratic politics which confines itself to a relatively narrow range of administrative activities without explicit acknowledgement or debate about the values implicit in the decisions that are made, or how those decisions relate to more fundamental questions about the nature and functioning of society (Phipps, 1990: 36). Contemporary politics appears more concerned with short-term adjustments among a limited range of possibilities, rather than offering qualitatively different visions of the direction in which society might develop.

This aridness of the political arena is profoundly evident in New Zealand where the two major parties give the appearance of advocating similar policies, as well as being equally contemptuous of the electorate to which they are responsible. This places democracy itself at risk. For, as Lasch (1991: 24) has written, the threat to democracy at present comes less from foreign totalitarian or collectivist movements than from the erosion of democracy's psychological, cultural, and spiritual foundations from within.

Politics may regain a degree of legitimacy in liberal democracies if it proves itself willing and able to address some of the substantial issues which face modern societies, such as the environmental crisis. In order to accomplish this task, however, there needs to be a clearer idea of the direction in which society should be developing. Instead of long-term decisions made within a framework shaped by a particular vision of New Zealand, decisions often appear inconsistent or contradictory, such as the withdrawal of public transport funding by the National Government at the same time as it claimed to be committed to reducing CO_2 emissions by 20 per cent by the year 2000.

Simply being critical of existing political processes and institutions is, however, not a sufficient basis upon which to build a coherent green programme. What is needed is a clear idea of goals and a credible strategy for achieving them.

Mobilizing Society's Resources towards Sustainability

Sustainable development must be the central component of future global development and should be an integral part of the long-term vision which New Zealand needs to develop. Politics in the era of modern society has facilitated unsustainable growth, from which it has been able to fund its programmes of social improvement. But now political systems must embrace sustainability as their overriding objective. To achieve this end, political institutions must take responsibility for mobilizing the immense energies of society—scientific, intellectual, economic, etc—towards achieving a sustainable society. This means each sector of the community, be it the agricultural or building industry, must create its own strategies for achieving sustainability. These sectoral policies could then be brought together in a national policy package representing New Zealand's response to the requirement for a sustainable society in the twenty-first century.

Too many of society's resources are currently employed in activities of dubious worth such as advertising and the arms industry. Green politics must push for sufficient resources to be made available so that everyone—from artist to engineer—who wishes to contribute their talents to the creation of a sustainable society has the opportunity to do so. This will mean very practical policies such as providing farmers who wish to switch to (sustainable) organic farming methods with transitional assistance. Engineers with an idea for water conserving technology should have access to the necessary loan capital to see their project brought to fruition, and obstacles to people using natural forms of health care should be removed. These kinds of specific suggestions amount to far less than the building of utopia, but the history of utopian experiments means that greens must be focused on what they can realistically hope to achieve.

Green politics embraces a critical view of the modern world, but it has to be recognized that this is a minority perspective. In a postmodernist future, it is quite possible that greens can create their utopian enclaves, and that green policies will become commonplace. Environmental values will become more common and will be incorporated into decision-making more and more. The anti-modernist emotions embodied by the greens are not likely to be universally embraced and, because they frequently impinge on the realm of personal space outside the gamut of liberal democratic politics, should not be the subject of concerted action by the State, although they will remain key elements of the green movement in its role as part of civil society.

Greens must, however, focus their energies towards the achievement of changing social goals and objectives so as to make sustainability the umbrella under which all other policies evolve. This will help to achieve the liberation of individual and social energies which are locked up by the assumption that people can do nothing about the state of the world. The 'green myth', identified by Blackwell and Seabrook (1988), is based on the view that 'the Apocalypse can be avoided by human effort, that a final ending of the human story can be prevented by our own insight and action' (Phipps, 1990: 65). It is precisely the responsibility of political institutions to facilitate this sense of empowerment, and action, beyond the current ghetto of green thinking. Green politics facilitates more conscious choice about the kind of future which people want by presenting people with different options and scenarios from those offered by conventional politics.

The role which green politics plays in providing alternative perspectives on the crisis of modern society is vitally important in helping individuals and society to adapt to a period of unprecedented change. Faced with the prospect of universal death and destruction, from environmental degradation and from nuclear holocaust, life and death are no longer issues which are faced solely by individuals, but face the human community as a whole. Survival issues must put all decisions made into a different perspective. Cousins (1981: 131) argues that 'nothing we do individually or collectively makes sense unless it is connected to the making of a structured peace'. The Holocaust and Hiroshima have shown the negative potential of modern society, obliging us all to speculate on the most profound metaphysical issues:

> Now we are all philosophers, in the sense that we all face the deepest issues of life and death, being and nothingness, issues that were previously seen as the preserve of an elitist group of academics. (Phipps, 1990: 59)

Every decision must now be made with cognizance of the fact that it is within human power to determine the future shape of all life on this planet. The smallest decision, or the largest, each contributes to the shape of the society we and future generations will inhabit. A single species now has immense (omnipotent) power to determine the shape of the future and, indeed, whether or not there will be a future at all. This 'humans as God' scenario demands new ways of thinking and behaving on the part of humanity. Yet our political institutions have been slow to take heed of this responsibility. In fact, as Suzuki (1990: 176) argues: 'our political system seems ill-designed to handle the major issues facing our leaders'. Anderson (1990: 169) reports that:

> A White House official told an acquaintance of mine that global environmental worries were generally regarded as "laughers" in the higher circles of government. Any staff person with a good feel for how to win approval and get ahead had sense enough to avoid bringing up such stuff . . .

The reform of our political institutions, local and global, and the election of competent greens to these bodies must be crucial goals of green politics.

Political institutions must prove their willingness to address matters of public concern, such as the overwhelming support in New Zealand for strong environmental protection measures (Gold and Webster, 1990: 44). The overwhelming vote for change to the electoral system in the electoral reform referendum in September 1992 showed just how strong is the desire for a change to New Zealand's political processes and institutions. In New Zealand, our political institutions are one of the few areas of life which have not been radically reformed in recent times. While most other aspects of life have been subject to the principles of competition and the 'free market', New Zealand remains encumbered with an unresponsive two-party system. Hopefully, this will soon change, posing new opportunities for the Green Party to present a coherent range of policies designed to move New Zealand in the direction of a sustainable future.

Green politics represents the hope that politics can move beyond the failure of socialism and capitalism, and '. . . to begin a new historic passage—towards a democracy that encompasses not only personal and political freedom, but the germinal decisions that determine how we and the planet will live' (Commoner, 1990: 176). There is a need for the evolution of the planet to be democratized and for people and nations to become partners in making conscious decisions now about what kind of planetary future there will be. Modern technologies, the emerging 'global culture' (Anderson, 1990), and events such as the Earth Summit make the conscious choice of future patterns of development feasible in a way that has been inconceivable to date.

The Need for Global Institutions

There is a need for individual states to resource and empower the international institutions necessary to deal with international environmental and developmental issues. Former New Zealand Prime Minister, Sir Geoffrey Palmer, has written of the need for 'a proper international environmental agency within the United

Nations which has real power and capacity' (Palmer, in *Victoria University of Wellington Research Report*, 1991: 38). In the same way that the welfare state deals with distributive issues on a national level, there is a need to address these issues globally, not just for reasons of justice but also because sustainable development may well depend upon it. Coombs (1990: 31) argues that the '. . . increasing scarcity of non-renewal resources will mean, both in terms of wealth and income, that most of the rich will grow richer and almost all the poor, poorer, will be as true internationally as it is within national boundaries'.

Global environmental issues will simply compound pressing issues of international development and the relationship of the 'developed' world to the 'developing' world. Not only do modern industrial nations consume a vastly disproportionate amount of the world's resources (approximately 80 per cent), but many of the human disasters in Third World nations are a result of ecological factors, such as the desertification of the African continent. Poverty is at the root of many problems such as deforestation and the depletion of soil and water, as well as the threats to endangered species.

Furthermore, the economies of Third World countries are often resource-based. The interrelationship between environment and development issues demands the creation and resourcing of institutions with the responsibility to develop and monitor solutions from a global perspective.

The Norwegian Prime Minister, Gro Harlem Brundtland, recently stated that the end of the Cold War provided an unparalleled opportunity for Western countries to redirect their military budgets to environmental clean-ups and assistance toward sustainable development in Third World countries (*Dominion*, 18 March 1992). Such calls have fallen on deaf ears in New Zealand. New Zealand has no environmental fund and, as a percentage of gross national product, its meagre overseas aid budget is reducing annually.

There are a panoply of global organizations ranging from transnational businesses to Greenpeace, some of which have greater access to information about environmental issues than smaller nation states (*Victoria University of Wellington Research Report*, 1991). But there are inadequate global institutions to respond to the issues raised by international non-governmental organizations (NGOs) like Greenpeace. NGOs have had varying degrees of success in their international efforts, while corporate actors tend to have varied interests with some actively working against international agreement on restricting climate- and ozone-damaging activities. The United Nations is the obvious body to take on greater administrative responsibilities through such bodies as the United Nations Environment Programme (UNEP). But this depends on the funding supplied by individual nation states, as well as on their political commitment to enforcing UNEP's policies, and the credibility of UNEP itself.

These problems of international governance have been highlighted by the outcome of the 1992 Earth Summit in Rio de Janeiro. The United Nations has no enforcement powers to back up the 'Agenda 21' which resulted from this vital gathering. There is a danger that the Agenda will fulfil pre-Earth Summit predictions that it might turn out to be 'a cosmetic solution not backed by political will' (*Time*, 30 March 1992). Green politics at every level must work to belie this pessimistic appraisal of the prospects for the Agenda 21, probably the most important agenda for sustainability yet produced. The impact of green politics in each nation will be one of the vital factors affecting that nation's attitude towards global environmental issues and the pursuit of sustainable development. There are signs of hope. The election of Clinton and Gore as American President and Vice-President, respectively, has certainly made the prospect of an American commitment to Agenda 21 brighter (*Evening Post*, 12 December 1992), which is important in a world where the Americans play such a role in resource consumption as well as in influencing the activities of the global market-place.

But in spite of some positive developments, administrative capabilities have not generally kept pace with the growth of global economic and cultural trends. As Anderson (1990: 251) argues: 'we are seeing now the emergence of a social reality that is different in important ways from anything we have known before, the first global civilization'. Postmodern civilization will be a global civilization, foreshadowed by international agreements such as the Montreal Protocol on reductions in ozone depleting gases. Other areas in which 'global environmental regimes' (Porter and Brown, 1991: 21) already exist include the protection of whales, international trade in endangered species and toxic waste, trans-boundary air pollution, the dumping of wastes and other materials in the sea, and the protection of Antarctica. These agreements recognize that environmental problems do not respect national boundaries. Institutions with monitoring and enforcement powers based on an awareness of the same reality now need to be established.

New Zealand was a leading proponent of the establishment of a United Nations with peace as its major focus. Could we not now take a similar role in promoting the establishment of bodies which fill the vacuum left in a situation where: 'The integrated and interdependent nature of the new challenges and issues contrasts sharply with the nature of the institutions that exist today' (Dryzek, 1987: 310)? Former Prime Minister, Sir Geoffrey Palmer (1990: 28), was a strong advocate of 'a conceptual leap forward in institutional terms':

> In New Zealand's judgement, the traditional response of international law— developing international legal standards in small incremental steps, each of which must subsequently be ratified by all countries—is no longer appropriate to deal with highly complex environmental problems of the future.

Green politics should be at the forefront of efforts to ensure that New Zealand's foreign policy includes taking immediate steps towards establishing appropriate global institutions to address today's global issues, including the achievement of sustainable development.

The Role of the State in Achieving Sustainable Society

The differences of opinion within green politics mean that there are a diversity of views about the best way for society to move towards the goal of a sustainable society. There is a spectrum ranging from free-market environmentalism—which claims that the free market is the nearest reflection of the decentralized, ecological principle (Anderson and Leal, 1991: 170)—to the 'eco-theologues': 'world savers for whom environmental requirements (as they define them) supersede the niceties of democracy' (Toffler, 1990: 312).

The State and its policies, as discussed in the previous chapter, must take a leading role if modern society is to adapt to becoming a sustainable society. Mishan captures this reality in the following statement:

> I do not believe that there is any escape from the dilemma facing us: if we are able to survive the perils posed by ecological hazards, by the permissive society and by the incipient computer revolution, it can only be at the cost of a more embracing and more repressive state. (Mishan, 1986: 192)

Whether a more embracing State need be more repressive is a matter of contention; the important point is that the achievement of a sustainable society will require a variety of measures, many of which will require an active role for the State.

Sustainability demands not just the kind of financial and legislative frameworks suggested in the previous chapter, but also new forms of administration. Green politics must propose and promote not only appropriate policies but also institutions designed to advance sustainable development and to address the environmental crisis: 'administration cannot simply be abolished: the historical possibility is for a form of administration more adequate to the environmental problems which are emerging as we move into the aftermath of industrialisation' (Paehlke and Torgerson, 1990: 291). The reform of existing policies will be an important first task. It will be crucial '. . . to reform the public policies that actively if unintentionally encourage deforestation, desertification, destruction of habitat and species, and decline of air and water quality' (MacNeill, 1990: 114). Policies such as taxation incentives to encourage the conversion of native forest to farmland—implemented in New Zealand in the past—should be the first to go.

Many public policies from the past, designed to address misdemeanours from a different era, now produce skewed outcomes and act as obstacles to innovation. For example, building codes originally designed to ensure that workers did not have to live in substandard dwellings, have been used to justify the demolition of homes built in alternative settlements on the South Island's West Coast. These homes might not feature the required two doors between toilet and kitchen, but they may offer examples of how resource-frugal and environmentally friendly dwellings can be built in future. Whereas the State's role in the past has been to uphold conformity and standardization, the State's role must now be to facilitate innovation through public policy, so that concrete steps towards sustainable development are encouraged.

Secondly, it has already been argued that the State must encourage innovation through redistributing the costs of developing environmentally friendly technologies and processes across industries. The State also has a role in addressing any social inequities arising out of moves towards a sustainable society, such as job losses in smokestack industries, or polluter-pays principles which will be a greater relative burden for those with less resources, where they are applied universally. The State will be in the best position to ensure that such inequities can be addressed through other policies, such as taxation measures, so that any socially adverse affects are mitigated.

Thirdly, the State also has a unique role in the protection of individual freedoms through constitutional (and other) mechanisms. Given a worst-case scenario of worsening environmental catastrophes and, for example, mass migration to escape rising temperatures or sea levels, democracy and accepted levels of individual freedom in modern societies could come under threat. Only the State has the power to ensure that even radical reforms are not inconsistent with the preservation of individual freedoms. The State is in the best position to make the decisions about the balance of individual rights versus collective benefits, for example when it comes to deciding between reduced CO_2 emissions and the rights of all people to unrestricted mobility.

For while environmental issues do not dictate a particular political programme, environmental issues inevitably involve political decisions and judgments. For example, rights to mobility are implicit in discussions about transport options, but rarely are different views about people's entitlement to mobility discussed. Yet given that mobility is inherently entropic (i.e., it uses energy) and that it is increasing mobility which is the cause of much pollution from roads, motor vehicles etc., it is inevitable that the question of people's rights to move and to travel must be explicitly discussed. Thus an environmental problem is easily transformed into a conventional political question about the rights of individuals against those of the collective, because the environment can be seen as the ultimate collective good. China's policy of one-child families is another interesting case in point. While representing an untenable infringement of individual rights in many people's eyes, it is an interesting example of the kind of choices which environmental demands (in this case, population control) bring to bear on policy choices.

Inevitably, given current institutional arrangements, it will be the State which makes these public policy choices. This perspective accepts the view of the State as a potentially neutral arbiter, but also, in the right hands, as a vehicle for essential reforms. This is not to deny the degree to which the State is dependent on capital accumulation, or other obstacles to reform. But there are simply no other extant institutions which can replace the role of the State in the move towards achieving a sustainable society. Even though 'the democratic logic is still to a great extent limited by, and subordinated to, the logic of industrialisation and capitalism' (Heller and Feher, 1988: 15), only the State has the potential to create a framework of sustainable development within which capital can operate. No amount of popular mobilization or active citizenry can deliver the long-term fiscal and legislative framework within which sustainability might be achieved, although popular movements will be crucial to promoting and legitimizing such policies.

Those adversely affected by the modifications of the market (such as those demanded by transition to a sustainable society), although they may be a minority of the population, often have 'disproportionate influence on public policy': 'The question, then, is whether the industrial democracies will be able to overcome political constraints on bending the market system toward long-term sustainability' (Ruckelshaus, 1990: 130–1). The power and will of the State will be central to the accomplishment of this task. Greenpeace and other grass-roots organizations will continue to have a vital role to play in raising issues which demand attention on the sustainability agenda, but it is difficult to envisage any institution other than the State actually putting in place the necessary policies to address those issues. Similarly, many businesses are now taking responsibility for putting in place corporate environmental policies, and taking practical steps through energy saving, recycling, etc. But these kinds of desirable practices will only become widespread when the State begins to lead the way in terms of making sustainability a central goal to which all sectors of the community must respond. Only once such a framework is in place will the economic disincentive for the majority of businesses to pursue sustainable activities be removed.

Financial penalties for behaviour deemed undesirable—such as those used by the Chinese government to discourage more than one child per family—have immediate effects on people's behaviour. Through the simple application of a measure like a carbon tax, which taxes carbon-producing products like fossil fuels, a whole new range of activities which were previously uncompetitive (because existing economic arrangements take no account of the environmental costs of fossil-fuel based activities) will become viable.

The very fact of internalizing the costs of air, water, noise, pollution, etc., which are currently 'externalized' by the market, will force changes in the viability of some existing industries at the same time as making the provision of new goods and services feasible. Regulations could be put in place which require steady improvements in the efficiency of technologies, including building standards and transport infrastructure, and a comprehensive resource accounting system would maintain detailed records of ecological stocks, showing where economic instruments might need to be employed to ensure a halt to practices which are depleting resources unsustainably.

Tomorrow's sunrise industries will be those which include the full social and environmental costs of their processes in their operating budgets. These issues are central to those being dealt with by the 'Cleaner Production' programme of the United Nations Environment Programme, which seeks to encourage business to take cradle-to-grave responsibility for its products and the processes which produce them. One of the outcomes of this kind of approach has been the development of a whole new field of 'green design' using novel methods and recycled products, etc. (Mackenzie, 1991).

Assisting the development of a new attitude from industry are examples from around the world of planning procedures which require a proactive approach to taking action before adverse environmental effects occur. Such an approach is usually backed up by legislation which permits heavy penalties for pollution when it does occur, thus discouraging industry from polluting in the first place. Some environmental legislation—in Washington State in the United States, for example—has already seen company directors imprisoned because they have been charged with personal liability for pollution. New Zealand's Resource Management Act 1991 includes similar, as yet untested, provisions. Such measures are essential if we are to change the attitude that nobody is responsible for collective resources, including air and water. But such provisions will only be made use of if sufficient green politicians are elected to ensure that the provisions are enforced.

Another important way in which producers will be encouraged to embrace a holistic perspective is for public budgets to be explicitly directed towards socially and environmentally friendly producers and products. The public purse is one player in the economy which has the power to send out important signals to the market-place, as the Australian Federal Labor Government recognized in 1991. The Australian Government launched an investigation into the complete federal purchasing budget in light of the desire to send out the signal to producers that environmentally friendly products would be the government's first purchasing choice. Some years earlier the Greater London Council (GLC) had also shown how the public purse could be used to encourage the production of socially useful goods, such as aids for the disabled and measures for improving the heating problems in council housing apartments. The GLC had also unsuccessfully attempted to implement a plan which was designed to encourage the reorientation of the London arms industry away from weapons manufacture and towards more constructive forms of production. A similar goal was articulated by Bill Clinton during the 1992 American presidential election campaign. His programme to achieve this objective is worthy of attention, for it attempts to address one of the most important issues facing the economies of many modern industrial societies where entire towns and regions may be dependent upon the arms industry.

Recognizing that technology cannot be treated as value-free and that there are social choices made by producers and designers, the GLC established guidelines for new technology so that it should meet 'social needs', which were defined as follows:

that in manufacture and in use it conserves energy and materials; that its manufacture and repair, and the recycling of its products can be carried out by non-alienating labour; and that its production and its products should assist human beings rather than maiming them. (Mackintosh and Wainwright, 1987: 204)

Acknowledging the importance of an active role for various levels of government in the pursuit of sustainable society, some writers have argued for a 'green social democracy'. Toffler (1990: 444) contends that Europe's main ideological export in the years ahead will be a 'green version of social democracy'. This is a useful indication of the potential shape of politics in modern industrial societies, and it suggests a desirable balance between private and public interests. It may well best encapsulate the argument which this book is advancing. But a green social democracy will only be of long term advantage if it challenges the shared social goals of both the political Left and Right, because: 'Political differences between right and left have by now been largely reduced to disagreements over policies designed to achieve comparable moral goals' (Lasch, 1991: 22). These comparable moral goals gravitate around the values of materialism and related (narrow) definitions of 'wealth' embraced by conventional economic thinking.

Towards a Redefinition of 'Economics'

The green desire to embrace a holistic perspective, to recognize the interconnectedness of seemingly disparate elements, is demonstrated nowhere more clearly than in the green approach to economics. Whereas modern economics tends to appraise even the most complex issues on a simple profit-loss basis, green politics argues that modern economics is value-ridden and reductionist. Aspects of decisions which may be the most important to the majority of people—for example, environmental protection or the quality of life—are often totally excluded from economic equations. Typically, economics has simply not devised the techniques for incorporating these elements into its equations. The urgent need is to develop a more inclusive definition of economics, as well as re-examining important economic concepts including wealth, risk, productivity, efficiency, needs, wants, and security (Robertson, 1990: 18).

The limitations of economics are exacerbated by the traditional indicators of economic 'wealth'. The typical indicator, the gross national product (GNP), increases with every car accident or with an environmental disaster, because these undesirable activities generate economic activity (car repairs, hospital services, the production of chemicals to disperse oil spills, etc.). This meant that Alaska's GNP increased dramatically as a result of the Exxon Valdez oil spill. A more holistic or 'inclusive' form of economics would clearly not exclude the social and environmental costs of these disasters. Nor would a holistic economics exclude the value of activities such as environmental improvements or domestic work, which go unmeasured by existing economic indicators.

There is a real need to develop indicators which provide a more balanced picture of the effects of various activities from social and environmental perspectives because GNP 'is a very narrow measure of total economic activity . . . [which] is a totally inadequate measure of the health of the economy or the wellbeing of the people' (Green Party, 1990: 3). Resource accounting and quality of life indicators are among the mechanisms which can be developed to give a more accurate picture of economic activity. Resource accounting, for example, enables the inclusion of the costs of unsustainable resource depletion in the accounts because it attributes a monetary value to long-term considerations in a way which contemporary economics does not.

The need for a redefinition of economics will be crucial to legitimizing many green policies, including the creation of measures to ensure that the market takes account of intangibles, such as the rights of future generations. The market has no mechanisms for taking the future into account, it cannot create social justice, it is powerless to make distinctions between good and bad, and it has no moral code (Seabrook, 1990: 188). This is not to say that the market is not a useful mechanism in certain situations, but it is to recognize that politicians must begin to show a 'willingness to subordinate the market to purposes that it is not geared to determine' (Daly and Cobb, 1989: 8).

The modern market has taken the place of traditional sources of morality, but it is based on an explicit set of values which assume that it is rational to pursue self-interest and personal accumulation. Behaviour which defies this, which is community-oriented or altruistic, is, therefore, deemed to be irrational. With the decline in traditional sources of morality, the market has become a primary influence on the way people lead their lives. The pursuit of 'progress', free of moral or ethical constraints, has encouraged the dominance of materialism and the pursuit of worldly goods as the main human activities in modern industrial societies. It is no coincidence that America, whose immigrant population was removed from conventional sources of morality and control in the Old World, was at the forefront of the growth of materialism and the dominance of monetary values. This is not to assert the need for a return to traditional sources of morality or control, as Lasch (1991), for example, would argue, but simply to acknowledge that the breakdown of traditional sources of morality has opened the way for the dominance of materialism and commercial values.

Given the minimal likelihood of any consensus surrounding new sources of morality, however, the need for practical political solutions to the problem of achieving sustainability comes to the fore again. The most important condition for sustainable development, MacNeill (1990: 118) asserts, is that 'environment and economics be merged in decision making'. Specifically, governments need to use economic instruments which encourage the raw material and energy content of products to be substantially reduced, as happened as a result of a concerted effort between government and industry in Japan after the 1970s oil shocks. Economic instruments are viewed unfavourably by some greens who argue that they entrench the rights of industry to pollute, or allow the wealthiest companies to pollute the most. But the possibility of combining economic instruments along with regulations and monitoring appears to be the most realizable solution currently on offer to address environmental concerns.

> The most effective anti-pollution programme is probably one that both has both rules and charges. The money raised by the charges can help pay for the rules. Not only does this help the government in times of tight budget constraints, but it implements the polluter–pays principle rather than merely pays lip service to it. (Meister in Broad, 1991: 19).

With this kind of policy package in place, in future it should be possible to create an efficient economy which provides more goods, services, and leisure time, while using less energy and materials, and producing less pollution (Salmon, 1991). The use of recycled materials can be encouraged far more widely, and a full cradle-to-grave responsibility enforced on manufacturers for their products. Including environmental costs in manufacturing processes will encourage less wastage, less energy consumption, less pollution, and will ensure safer and healthier working conditions for workers. A national budget which levies taxes on energy, resource usage, and pollution will send definite signals to the market, at the same time as encouraging new consumer behaviour (by increasing the cost of socially and environmentally less desirable goods). New Zealand's Resource Management Act makes these kinds of pricing mechanisms and policies feasible to an unprecedented degree, but a political commitment will be necessary to implement such measures. Again, a clear role for green politics emerges.

Redefining Meaningful Lives

The current environmental crisis is not the only catalyst driving the need for new directions in modern, industrial societies. Economic malfunctioning, leading to the creation of a permanent minority of people without access to the rewards of conventional employment, demands a rethinking of modern society's goals. 'Mass unemployment' suggests the need for redefining the notion of 'work' as well as the composition of a meaningful life. Economic security has underscored the emergence of the 'new politics', but economic insecurity also raises fundamental issues about social goals and values. Given the decline of 'full employment', society must establish a new goal of providing all people with 'full lives': lives in which paid employment may be only a small part of a person's time, but lives which ensure that every person has the opportunity to develop their skills and to contribute to broader society in ways that are meaningful to them. These broad humanist objectives must be an integral part of the green politics agenda.

Employment, work, and time issues are connected to the need to redefine economics. That which we currently define as 'the economy' is really the industrial economy, which barely acknowledges the vast variety of human activities that Berry (in Daly and Cobb, 1989: 18) describes as comprising the 'Great Economy'. Green politics projects new moral criteria for appraising the decisions of government, but green criteria also extends to judging the desirability or otherwise of various forms of human activity. We must begin by defining what are intrinsically worthwhile human activities rather than only valuing activities for which someone else—i.e., an employer—is prepared to pay. This is the great challenge which the current mass unemployment presents modern society with: rediscovering activities which are personally and socially beneficial even when the traditional 'job market' does not recognize such activities by providing payment and/or status.

Policy-making must now take into account the fact that it is highly unlikely that modern societies will again face a situation where the majority of people are employed in what we have conventionally come to call 'work': that is tasks performed in the paid workforce outside of the home. Exactly the same job, perhaps building a set of cupboards or caring for children, performed at home on the weekend, is not considered to be 'work', even though it does not differ from the same task performed in a different environment and clearly recognized and rewarded as 'work' during the 'working week'. These unpaid tasks comprise the major part of the informal economy which is already a significant and growing part of most modern economies. The informal economy comprises households and neighbourhoods 'where we and our families and our neighbours provide ourselves and one another with useful and necessary goods and services, for most of which no money changes hands' (Robertson, 1990: 30).

An essential role for green politics is to ask questions about how people in a postmodern future might realize their human potential even where they do not have access to conventional paid employment. The Greens in the European Parliament have published extensive reports on the move to a culture-oriented, as distinct from a work-oriented, society. New forms of socially sanctioned activity (and inactivity) might take the form of greater time for leisure or contemplation, or it may mean massive environmental reparation projects or encouraging individuals to turn their passions and interests into careers through provisions such as a guaranteed minimum income. Many people may forgo their conventional jobs to take up writing or other creative and artistic pursuits were it not for the economic necessity of paid employment. A guaranteed minimum income would rectify this situation by providing a regular payment, sufficient to live on, as of right to every adult citizen. There would be an incentive to supplement what would be a minimal income, but for people who wished to pursue creative or voluntary activities it would remove financial insecurity and allow them to pursue their interests to the full. It would also challenge the power of employers who would need to make their work and working conditions attractive and flexible in order to recruit workers. Such provision would obviously require increased taxation, but must be an option worth serious consideration as part of a package to address the current employment situation.

The goals of beauty and aesthetic sensitivity are important components of the green movement, with implications for redefining meaningful human activities. The Values Party's early publications frequently referred to the desire for a 'beautiful world' (Levine, 1975). This objective alone would validate a wide range of activities and pursuits currently undervalued by a production-oriented system. Marcuse (in Martineau, 1986) asserts that the age of relative material plenty in most modern industrial societies demands a rethinking of social goals, away from quantitative to more qualitative goals, among which he lists aesthetics as the most important. The goal of beauty legitimizes a whole new range of human activities, particularly in the cultural and artistic spheres.

In the United States, one million 'environmental jobs' were created between 1970 and 1980, making this a rare growth sector in the job market (Miller, 1991: 270). No detailed research has been done on the number of 'environmental jobs' in New Zealand, but the need to reafforest large parts of the East Coast of the North Island, for example, suggests this is a potential area of employment growth. Recycling now employs over a thousand people throughout New Zealand. In Wellington, employment creation groups like New Hope Recycling are meeting social and environmental needs with labour-intensive recycling schemes. Proper attention to the panoply of alternative services and ventures which are emerging at a local level would reveal that many jobs are being formed in novel and innovative ways, many outside the 'formal' economy, but all providing meaningful activities for individuals who might otherwise find the need to 'kill time' (an expression of modern society which demonstrates the degree to which time has become a burden to be dealt with, rather than an endowment to be enjoyed).

An important part of many activities in the 'informal' and 'alternative' sectors is their recognition that work is only one component in people's lives. People's non-work roles, within the family and the community, and with lovers and friends, are usually as important as their work and should not therefore be sacrificed on the altar of the forty-hour week. For those who still have access to conventional employment, more healthy work environments and power-sharing in the workplace will enhance decision-making as well as the awareness of the social and environmental consequences of corporate activities. For those wanting to set up in business alone, or to work from home, the necessary capital and/or technology should be freely available in order to facilitate the autonomous lifestyle preferred by growing numbers of people.

In short, the current employment 'crisis' is potentially a time of opportunity. Green politics can, like the established parties, perpetuate the myth that full employment will return again, or it can use this opportunity to generate a wide-ranging discussion about the need for 'full lives', rather than 'full employment'. The redistribution of existing work through measures such as the introduction of a thirty-five-hour working week, and other innovations designed to give all people access to a meaningful life, must be addressed in New Zealand.

Alternative Ways of Living

In pursuit of meaningful lives, increasing numbers of people seek alternative ways of living (or 'lifestyles'). Alternative ways of living are far more widespread than is commonly realized. Whether several inner-city houses share living facilities in pursuit of a greater sense of community, or a larger number of people choose to live in a community in an isolated part of New Zealand, many people are now making more conscious choices about their lifestyles. Conscious lifestyles and intentional communities are designed to enable people to live in ways which are more attuned to their personal feelings and preferences. Alternative lifestyles often attach greater importance to the sense of community, as well as providing people with the chance to live with people who share the same values. They represent the fact that 'ecologically sound personal lifestyles and practices' are one component of an ideal green society (Pepper, 1991: 8).

Alternative lifestyles and intentional communities also have a broader social and political significance. For example, they contribute to an 'alternative society strategy of change' which focuses on the promotion of new and different ways of doing things and an alternative culture, which contrasts with existing social arrangements. Coombs (1990: 13) says a major goal of sustainable development will be the provision of the intellectual, cultural, and social components of our desired lifestyles; making the primary indicator of economic well-being 'the number of people with access to healthy, stimulating, and dignified lifestyles' (Coombs, 1990: 13).

Alternative lifestyles have been practised in New Zealand by many people since the 1960s, and even beforehand by, for example, the Riverside Community near Motueka. Originally founded by pacifists during the Second World War, Riverside Community is one of the longest-surviving intentional communities in the world. An early member of Riverside stated that the '. . . Community enables to some extent the liberation and expansion of the personality, the stretching of one's capacities and powers, and fulfilment in ways that individual living cannot' (Cole in Rain, 1991: 197). Other communities have been less long-lived. These include the majority of the previously mentioned ohu of the early 1970s. Ohu were the idea of Norman Kirk (Labour Prime Minister, 1972–4) who was inspired by the Israeli kibbutzim. Kirk 'saw a kibbutz-type environment as an antidote to the ills of modern society, as well as a means of showing the virtues of a simpler life' (Hayward, 1981: 173).

There are numerous successful examples of alternative ways of living in New Zealand, and many Māori have maintained what the dominant system perceives of as 'alternative ways of living'. The Tui Community in Golden Bay in the South Island is another successful example of a conscious choice of an alternative way of life. This intentional community is based around an old homestead and offers a communal lifestyle as well as employment through industries such as bee balm production. The community's statement of intent includes the commitment to 'help in the regeneration of our planet and to create a sustainable future which can give our children hope'. The community follows no specific political creed or spiritual leader but 'a personal wish is to become whole and happy people'. The community encourages tolerance and understanding of differences 'through experiencing life together, we have come to some basic agreements . . . which set a tone for community life'. Interestingly, many of the permanent members of the community are foreign-born. Their community's creed is an explicit example of the connection between personal growth, lifestyle, and the planetary future. Not all attempts at communal living have been as successful as the Tui Community, nor do the majority of people choose to live in this way. In New Zealand there is ample opportunity for people to experience communal lifestyles in places like the Coromandel Peninsula and Golden Bay, as well as in some shared urban households. The fact that most people choose conventional lifestyles should not be overlooked by those who advocate the need to return to collectivist communities as part of a green social order.

Temporal and Spatial Rights

The Politics of Time

Increased self-realization and the conscious choice of lifestyles are likely features of a postmodern world. But such choices will be limited by the way in which time is currently demarcated. It is therefore likely that a postmodern future will also include a new attitude towards time. Challenging conventional notions of time already underscores many green approaches to lifestyle and work, etc. This heightened awareness of time, and its social control, is another example of the 'new politics' agenda's attempts to broaden the scope of political debate in modern industrial societies (Maier, 1987).

Giddens (1990: 178) argues that a postmodern world will likely include a radically different organization of time and space. Nowhere is the anti-modernist, anti-industrial movement thrust of green politics more applicable than in the need to escape from the dictates of modern industrial society's demarcation of time. Many modern political struggles gravitate around the desire for people to gain greater control over their time. Deutsch (1985: 18) notes that the growth of green politics in a European context was directly linked to the fact that: 'Imagination, spontaneity, affection in small groups—all these are now urgent human needs'. Yet the temporal organization of modern industrial society militates against the meeting of these basic needs.

A variety of movements have arisen directly seeking to challenge industrial society's discouragement of spontaneity and to give people control back over their time. Gorz (1989: 28) notes that 'One of the crucial (if less noticed) themes of contemporary . . . politics is the need for a new politics of time'. These new demands for temporal and spatial rights have arisen out of the scarcity of time in modern society. As Melucci argues 'The need to establish a new rapport between inner time and social time creates a demand for reversible time, for autonomously chosen and regulated units of duration unburdened by the rhythms of clocks and calendars' (Melucci, 1989: 178). A new politics of time will also be essential to developing constructive responses to the end of 'full employment' in industrial societies.

Modern industrial society's attitude to time has had two particularly adverse effects: one is to equate time with money, the other is that time has effectively been speeded up. The beauty of speed was worshipped by the fascistic Italian Futurists in the early decades of this century, while the cultural doyen of the Bolshevik revolution, Mayakovsky, declared: 'He who is without a watch is not truly a man'. Left and Right both recognized the importance of new attitudes to time and the magic of speed. Changing the organization of time was crucial to the establishment of new political orders. This demonstrates the extent to which time and its use, control, and delineation are inherently political.

Some writers argue that time-saving and 'freeing' devices such as the motor car, actually trap us in a cycle of long work hours to maintain the goods which we supposedly buy to increase our autonomy. Thoreau discovered at Walden Pond that he could live on six weeks' labour per year (Shrader-Frechette, 1981: 178). The minimal work hours of some forms of pre-industrial social organization are now viewed romantically by critics of modern work practices.

During the 1992 electricity crisis in New Zealand, one of the suggested ways to reduce energy usage was to reduce the working week to four days (*Evening Post*, 12 June 1992). This demonstrates that the way we use our time is not only relevant from the perspective of personal autonomy, but also relates strongly to patterns of production and consumption:

> The consumption traps . . . are just the flip side of the bias toward long hours embedded in the production system. We are not merely caught in a pattern of spend-and-spend —the problem identified by many critics of consumer culture. The whole story is that we work, and spend, and work and spend some more' (Schor, 1991: 125).

Arguably, the introduction of a four-day or thirty-five-hour working week could be the single biggest step towards a greener society which we could take as a nation. As well as reducing the amount of production and consumption, it would free people to spend more time in activities enabling self-sufficiency, i.e., growing food and mending goods. It would also allow more people time to contribute to the altruistic economy which already exists through voluntary community work, and give people time to develop new skills and interests. Care of the young and of the elderly, caring for the environment, working for peace, and giving service to Third World countries are all activities which more people would have the opportunity to participate in if we had a more flexible attitude towards working life. Such proposals might be called uneconomic, because they would reduce the amount of conventional economic activity, but the fact that a four-day working week was even considered in response to an emergency situation shows that it is a feasible option for New Zealand. And it is precisely the kind of option which greens must embrace as an example of combining social and environmental objectives while addressing the crucial issue of how the use of time is going to be reconsidered in postmodern society.

Green Cities

As Maier (1987) has argued, new politics raises new issues not just about temporal rights, but also about spatial ones. In terms of spatial rights, the city is vitally important. But the city is also central to the creation of a sustainable society, which is one of the main reasons the city must become a major focus of green attention in future. Cities are the largest users of energy and resources and the largest concentrations of pollution, etc. Because of this, they have an important impact on the natural environment far beyond their boundaries. The city cannot be separated from the countryside. Urban people's lifestyles, their consumption patterns, their mobility, etc. all have an impact on the 'wilderness', whether through the encroachment of refuse disposal sites on urban hinterlands or the need to temper wilderness with technology to generate energy for the cities.

Wilderness concerns have dominated the green movement in former frontier societies like New Zealand (Eckersley, 1990: 70). This has meant that green politics, particularly in an antipodean setting, has paid too little attention to the city. Yet the majority of people in modern industrial societies live in cities. The city is the arena of most individuals' daily experience and the environment which most shapes their lives. The city acts as the primary generator of socio-economic change in modern societies, a key fact when the creation of a sustainable society is being promoted. Recent examples of unsustainable economic activity in New Zealand have been based in the city. Speculative investment in urban real estate was the engine of the unsustainable mid-1980s economic boom in New Zealand which destroyed much of New Zealand's built heritage and replaced it with a vast over-supply of low-quality office space built by speculators with little commitment to the places where they undertook their activities.

But as well as being the site of many misdeeds, the city is also the potential source of many solutions to contemporary problems, given that cities usually harbour the multiplicity of new movements which promote innovative solutions to the problems of modern society. Economically, socially, culturally, and politically, any political strategy which is serious about change must address the politics of the city, even when it secretly harbours romantic notions about everyone returning to the land and abandoning the decadence and alienation of the city. The city and how to make it environmentally sustainable and truly liveable must be a primary focus of the contemporary green movement.

Urban development is, after all, inextricably connected to finding solutions to global environmental problems. For example, urban land use patterns are linked with resource depletion. Low density cities (like Auckland) encourage a greater dependence on private cars, whereas spatial concentration (as in Wellington city) makes the use of public transport and bicycles more viable. High density and integrated land use leads to a conservation of resources and to a sense of community which encourages the social interaction which is a major strength of urban living. 'Land use apartheid' should be replaced by greater mixed-use planning, where the emphasis is on the environmental effects of an activity, rather than on the regimented separation of activities (leading to greater car use as people are forced to travel long distances between work and home). These principles have long been adopted in European cities. Their application to an antipodean setting where greenfield development still consumes large tracts of farm land may be made easier by the Resource Management Act 1991, which presents the potential for a more enlightened approach towards subdivision. The Australian Federal Government has recently made urban consolidation a major policy plank in an attempt to reduce the destruction of urban hinterlands and the waste associated with new infrastructural development. No similar exercise has even been suggested for New Zealand's cities.

Scandinavian nations tend towards a policy of decentralized concentration in their urban planning. This encourages the development of activities around decentralized nodes based on public transport terminals. This takes the pressure off escalating property prices in central city areas, with the effect that buildings are not prematurely demolished to make way for taller structures. This reduces the number of high-rise buildings, as well as the pressure for business to encroach into residential areas that adjoin the city centre. But the development of this kind of urban development pattern presupposes a strong planning system, a presumption not always valid in New Zealand where planners have too often been the handmaidens of short-sighted development.

Town planning and architecture directly affect the lives of the majority of people in modern industrial societies. In recent times, town planning has come to see itself as a facilitator of investment and development, rather than as a discipline with a forward-planning role and a commitment to upholding the public good against the self-interest of property developers. Perhaps it is time to suggest a new focus for architects and planners: to make the city sustainable. A major focus of the redefinition of meaningful activity must be concerned with making the modern city more liveable, more environmentally sustainable, more green, and more beautiful. There is a growing body of literature on the 'greening of the city', dealing with everything from the planting of urban forests and farms, to encouraging building design which minimizes energy use, employs sustainable sources of building materials, and ensures a long life-span for built structures. Planning must reflect the changes in society more generally. A postindustrial land use policy will blur the distinctions between town and country, enabling a greater integration of home and work, and the possibility of creating greater self-sufficiency in the city in terms of areas like food production. The concept of 'quality of life planning' has also been suggested, and has been successfully applied to rescue declining cities with a focus on creating beautiful, green, and entertaining conurbations (Nicholson-Lord, 1987: 171–2).

Corporate status has been more important than any sense of commitment to the local environment in determining the shape of our cities and the modern architecture which dominates them. Architects have concentrated on serving the client rather than on the end-user. This does not have to be so. In Sweden, for example, workers have to be consulted about the design of their workplaces. If an architect does not agree to meet his/her client's demands in a New Zealand context, however, there will always be another more compliant architect available. Architects themselves have been described as 'a modern priesthood; entry is given only after initiation rites, and critics are damned for their audacity or patronized with mild contempt' (Short, 1989: 42). As Short (1989: 43) says, the design of our cities is far too important to be left to architects alone because: 'Our cities have become the graveyards of outdated architectural theories. The giant glass towers, once the buildings of the future, are now the tombstones of architectural modernism'.

Maintaining our built heritage has already been suggested as a major source of potential employment and economic activity, but there are other good reasons for preserving and enhancing our existing building stock. From a resource point of view, we cannot afford to rebuild our cities every generation. The sheer waste of resources makes this prohibitive. Frequent building is also socially and functionally disturbing. There are also psychological reasons. City dwellers value familiar surroundings and the intrinsic value of the sense of place which has so frequently been destroyed in the name of progress. People feel dominated by high-rise buildings, not to mention the shading effects and wind tunnels created by high-rise buildings. Modern high-rises also tend to be air-conditioned and central-heated, consuming vast amounts of energy just to ensure almost uniform discomfort. As a society, we have decided that it is better that everyone should be supplied a uniform amount of air and light, rather than develop methods of resolving disputes between one person who wants a window open and another who wants the window closed. Sick-building syndrome is the result, with a loss of working days especially as sealed buildings fill up with modern technology which subject workers to a variety of gases and fumes.

A sustainable city will maximize the use of natural light and air in its buildings (which makes good sense economically, in reduced energy bills). As it is, approximately 50 per cent of the heat generated by modern office buildings escapes out the roof and contributes to global warming. But high-rise office buildings are also related to particular patterns of working life, reflected in the spatial layout of cities concentrated around a 'central business district' (more accurately, as Melbourne city has recognized, a 'central activities district'). People travel from the suburbs into the inner city to work for a set number of hours per day, and then return home to their 'dormitory suburbs', so-called because they are used for little other than to provide beds for weary workers. In the technological age where it is possible for people to work from home via a computer connected to a central workplace, or where firms can locate in cheaper locations and maintain business activity through technological links over long distances, the need for huge concentrations of people in inner city areas is reduced.

Urban sustainability should be pursued for other than purely environmental reasons. As mentioned earlier, in New Zealand during the mid-1980s there was a property boom created by the opening up of money markets and the availability of large foreign loans to local borrowers. Much of this money was speculatively invested in central city real estate. The effects on New Zealand cities have been devastating, with the loss of much of our built history and the subsequent erection of characterless, glass-clad boxes which say nothing distinctive about New Zealand as a nation, except perhaps to reflect the sad fact that aesthetics are still not highly valued.

It is in the city that growth is primarily sold as a 'public good'. Everyone, it is alleged, benefits from growth and from the increase in economic activity as a result of 'development'. Growth brings jobs, increases the city's revenue take from sources such as rates and land tax, and increases the size, and therefore the status, of the city in relation to other competing conurbations. City government is seen to be reneging on its responsibilities if it fails to attract investment, with the resultant growth. In support of this is an interconnected network of pro-growth groups, from the public and private sectors, united behind a commitment to a doctrine of value-free development in what has been described as the 'growth machine' (Logan and Molotch, 1987: 32).

But urban growth is not necessarily beneficial to everybody. Logan and Molotch (1987) assert that: 'we are certain that local economic growth does not necessarily promote the public good . . . development projects that increase the scale of cities and alter their spatial relations inevitably affect the distribution of life chances', while Short (1989: 122) claims that: 'We pursue growth to the detriment of our collective health, ecological resources, and long-term sustainability'. Local growth often creates no new employment, but simply redeploys existing workers and redistributes current jobs. The benefits of growth need to be more accurately assessed in terms of their effects on particular groups and particular places. To restate a recurring theme, the costs of development also need to be clearly enunciated. A new subdivision, for example, may cost vast amounts in terms of supplying services such as water and sewage to a small number of new homes. The amount spent on these services for a few will be a cost to the rest of the community who may have preferred that the expenditure be directed towards the improvement of existing services and towards enabling a higher density living in existing residential areas. The destruction of a green field or an open space environment for new homes may remove important recreational facilities for poorer people who depend more on public open spaces. Logan and Molotch (1987: 98) conclude their investigation on local growth with the salutary warning that:

> In many cases, probably in most, additional local growth under current arrangements is a transfer of wealth and life chances from the general public to the rentier groups and their associates. . . . To question the wisdom of growth for any specific locality is to threaten a benefit transfer and interests of those who gain from it.

To challenge the assumption of the unequivocal benefits of growth in the city is to question the assumption of the desirability of growth throughout society.

In the 1960s and 1970s, patterns of urban development were frequently at the centre of cultural critiques in Britain and Europe, as motorways and commercial developments transformed inner cities, displacing long-established residents and counter-cultural activities attracted by low rents. An urban counter-culture emerged, with its primary target being the bureaucratic local administrations and their programmes of urban renewal. The urban opposition was characterized as an 'investment obstruction' but it was the source of many creative ideas, including an involvement in the establishment of experimental ways of urban living such as Christiania in Copenhagen. These local groups also acted as the genesis of green groups which contested local elections and eventually gained institutional representation in their battles against what were perceived to be undesirable development and planning practices. Taken together, the accumulation of local crises were dramatic and led to the increased attention to local politics and calls for its modernization (Roth, 1991: 81). This posed a challenge to the traditional State-oriented attention of politics, especially where social democratic regimes had strengthened centralization. The inability of established parties to capitalize on the emergence of new local movements was one of the reasons for the subsequent emergence of green politics, which included under its auspices a wide variety of movements and concerns.

In New Zealand, the Values Party was active in campaigning against the high-rise BNZ headquarters in Wellington, as well as demonstrating against the urban motorway. But urban issues remained relatively insignificant when seen alongside wilderness issues such as the felling of forests and the damming of rivers. Observing green politics in New Zealand, one could be forgiven at times for thinking that this was not a highly urbanized society. But New Zealand is, in fact, a highly urbanized society, and depopulating the cities and encouraging people back to the land is neither environmentally sustainable nor socially viable. Depopulating the cities has been the hallmark of repressive regimes like Pol Pot's

Kampuchea, which emptied the cities and forced people to live subsistence lifestyles in the countryside (resulting in a massive loss of life). People are likely to want urban living for the foreseeable future, and the possibility for them to leave the city already exists. The challenge for green politics is to recognize the centrality of the city to the green agenda and to apply the concept of a sustainable society to the urban context.

Conclusion

The great challenge for a modern society is to channel its individual and social energies into achieving a sustainable society. Unprecedented access to information and a variety of skills—from 'envisioning' processes to modern strategic planning and management techniques—mean that there is an unparalleled opportunity for societies to make informed choices about their future directions. Similar opportunities now present themselves to increasing numbers of individuals through the personal growth and New Age movements. These processes mean that decision-making can take place with a long-term perspective which encourages sustainable human activities and takes future generations into account '. . . so that the very thought of destroying the world becomes utterly inconceivable' (Phipps, 1990: 77).

The State will be fundamental to the achievement of green reform, for it remains one of the few institutions to match the power of private capital, and has substantial powers in determining national goals and priorities. The nation state retains the power to enforce regulations and to implement economic packages designed to encourage particular types of behaviour deemed desirable by society. A stronger role for the State is not incompatible with pluralism: the State has the potential to create a decision-making environment which will encourage sustainable development, without repression or conventional control mechanisms such as the ownership of key resources. No other body currently has this power to delineate the frameworks within which future development must occur. There is a need for analogous institutions on a global level, with administrative bodies based on regions and localities playing a crucial complementary role. The role of the city in the achievement of the sustainable society is vital and must not be overlooked, in spite of green ambivalence towards the urban conurbation which is the home of the majority of people in modern societies.

The ability of political institutions to deal effectively with the demands of the green agenda will be vital to ensuring the continued legitimacy of liberal democratic political systems. Just as the social democratic and labour parties have changed social objectives over the last century—making the plight of workers visible in the production process—green politics makes visible factors which are currently externalized, such as environmental and quality of life considerations. An awareness of these kinds of issues is demanded not only in the light of the global environmental crisis, but also because of the end of full employment, and the consequent need to redefine meaningful human lives.

Radical new approaches to the definition, arrangement, and distribution of 'work' must be a priority for green politics. Rejuvenating the goal of providing all New Zealanders with the conditions for a happy and fulfilling life will do much to create the conditions where people are able to grow and to increase their ability to identify with others and with the 'natural' world.

Green politics requires that society should look afresh at the issue of time and how it has been commandeered from the majority of people and placed in the hands of employers. Demands surrounding time and personal autonomy are critical in many contemporary political issues and form an integral part of the green agenda. The same sense of responsibility which people now wish to take for their time needs to be applied to decisions affecting the future of global evolution. What is demanded now are appropriate forms of governance which facilitate conscious human choices about the future direction of the planet on the basis of an awareness of human responsibility for global evolution.

5 The Green Contribution to a Postmodern Future

The vision of a new system of transparent power replacing the old and unjust system has claimed too many victims in our century. At the very least, respect for these victims obliges us not to feed such grandiose and dangerous illusions any longer. (Melucci, 1989: 189)

In this chapter, the green ideology and suggested agenda will be placed in the context of the likely factors shaping the future of societies like New Zealand. For while it is easy to espouse all manner of romantic visions about how society ought to look, in the end politics is the art of the possible. Unless green politics accepts this premise, a repeat of the same sense of disillusionment and hopelessness that eventually destroyed the Values Party appears likely. An analysis of how modern societies are likely to evolve therefore becomes vital.

As we have discussed above, the green rejection of modernism leads to a variety of responses, ranging from romanticizing about the pre-industrial past to a belief in the imminent arrival of a qualitatively different postindustrial world. If it is to be effective, green politics must reject any illusions about the inevitability of achieving a full-scale social transformation (or utopia). Instead of dissipating green energies on the expectation of a 'new paradigm' or a 'New Age', greens should be focusing on the steps that they can take now to redress the unbalanced developmental path down which modern society has careered.

In undertaking this task, greens can learn from the many concrete pieces of a green future which already exist. Alternative communities, organic farms, and natural healing centres provide the foundations upon which the green contribution to a postmodern future can be built. Each living, working example of sustainable activity is a step in the desired direction; tangible evidence of how a greener future might look.

Greens might also learn a valuable strategic lesson from the concrete utopias which already exist throughout New Zealand: that is, everybody has the choice to join in such 'alternative' activities and ways of life, but few people actually do. People may participate in particular alternative activities—going to a natural healer rather than a doctor, or gardening organically—but there is little evidence that people want to fundamentally change their entire lifestyles. Therefore it is unlikely that a generalized 'green' transformation, even if it were achievable, would gain widespread support, particularly given the sacrifices it would entail. Any proposed transformation must confront the issue of what to do with those people who do not want to live in a qualitatively different way, as envisaged by the green utopia. This issue has been inadequately addressed by the proponents of a green transformation.

The alternative to the potentially totalitarian green utopia is a package of practical policies and reforms. Such a plank inspires none of the fervour engendered by the hope for a utopian transformation, but does offer the most likely path towards a sustainable society. It also offers the best guarantee of reform which is compatible with the protection and enhancement of human rights. Let us look at the environmental utopia in a little more detail to confirm this argument.

Ecotopia

Callenbach (1975) entitled his novel of an environmentally based utopia, *Ecotopia*. Ecotopia is a totalizing project based on the dark green 'fundamentalist' agenda. It encapsulates the likely shape of a society governed by the values of Lee's (1989) 'ascetic socialism' (discussed above) and other green utopias. Ecotopia is reminiscent of other 'mass' solutions from the industrial age in that there is no room for individuals within its all-encompassing grand scheme. Ecotopia embodies a break in institutional continuity, a new model of society arising in response to the crisis of modern, industrial society. Ectopia satisfies Bahro's assertion that 'The problem is not to create a space for minorities but to create a new solution for the whole of society' (Bahro, 1984: 218).

Ecotopia is a society based on ecological principles. For those at the fundamentalist end of the green spectrum, 'ecology' has taken on the guise of a single truth by which all actions might be judged. For some greens, ecology has become the new absolute. To this extent, it is analogous to the absolutism of a vulgar Marxism, although religious fundamentalism may be a more relevant parallel in the contemporary world. Both movements reject modern society, seeking the establishment of a society based on their own preferred set of unequivocal and absolute values. Greens who adhere to this 'deterministic' outlook tend to believe that all must be sacrificed in order to avoid the inevitable apocalypse. In their desire to re-enchant and again make sacred the world, those described by Toffler (1990: 398) as 'eco-theologues' are committed to restoring a 'religion-drenched world that has not existed in the West since the Middle Ages'. They share with religious fundamentalists a deep hostility to secular democracy and threaten to return the world to a pre-industrial order based on asceticism and self-sufficiency. Ecotopia rejects modernity in its entirety. The pre-industrial past becomes the primary indicator of ideal social arrangements for the future. Green fundamentalism embraces none of the tolerance of diversity which underlies democratic thought.

Callenbach's ecotopia is a State of closed borders with no air travel allowed even over its territory. No one litters in ecotopia, and its citizens know the name of every example of flora and fauna they see. Citizens exhibit social characteristics of patience, openness, and communality, qualities not seen in the countries beyond ecotopia's borders. The emphasis of such a State is upon self-sufficiency and respect for the environmental frontiers and limits within which it resides.

Ecotopian schemes have gained intellectual legitimacy in parts of the contemporary green movement through concepts such as the 'bio-region'. 'Bio-regionalism' embodies the idea that the size and form of human communities should be confined by the natural environment of which they are a part. The bio-region 'is an area whose boundaries are determined by natural rather than human dictates, distinguishable from other areas by attributes of flora, fauna, water, climate, soils, and land-forms, and the human settlements and cultures whose attributes these have given rise to' (Sale, in Davis, 1991: 8). Bio-regionalism looks favourably upon past eras when tribal groups had strong associations with particular places which provided them with their history and culture. Sale argues (in Anderson, 1990: 247) that 'to come to know the earth fully and honestly, the crucial and perhaps only all-encompassing task is to understand *place*, the immediate specific place where we live'. The campaigns for separate States by Basque, Welsh, or various Slavic nationalist groups, are seen to be manifestations of this desire for a political map made up of 'bio-regions' rather than 'artificial' nation states. Many cities, regions, and nation states already reflect bio-regional boundaries, but the emphasis of bio-regionalism is on limiting human populations and on ensuring that human activities recognize the constraints placed on them by the natural environment.

A recent article on bio-regionalism in a New Zealand context suggested that the Canterbury Plains or the South Island's West Coast would be potential bioregions (Davis, 1991). Such boundaries accord with the regional governments which governed colonial New Zealand, as well as reflecting New Zealand's current regional council boundaries. Within each bio-region, Davis (1991: 8) asserts, 'strength is drawn from belonging to a particular region and the identity that stems from intimate historical and spiritual connections with that area and its history. The key to this strength is the people's reverence for the land which they hold sacred.' Within the bio-region will be a 'stable state' economy which minimizes resource use and emphasizes waste reduction and recycling. Economic activity would be confined to that which can be undertaken with local raw materials and trading would be restricted across bio-regional boundaries. There would be a heavy emphasis on local self-determination, with a voting system which reflected the acreage of the bio-region as well as the human population. Cities over 50,000 would be 'too big', and a multiplicity of small cities would 'mean that the city dweller has ready access to the spirit of nature that is embodied in the countryside' (Davis, 1991: 9). This form of social organization would come about not through revolution but through 'a gradual process of education, reshaping and recreating—an evolution. It does not require major leaps of social patterns, rather supporting the inherent needs that run through everybody' (Davis, 1991: 9).

This ecotopian vision of radical decentralization and increased powers at a local level would involve a radically different way of life. There would be a far greater emphasis on self-sufficiency in daily living, and less focus on material goods and the work patterns necessary to sustain a high-consumption society. Artificial stimulants, such as coffee and alcohol, and passive or high-tech forms of entertainment would give way to simpler pleasures and active leisure-time activities possibly related to self-sufficiency or to the preservation of traditional cultural patterns. This model reflects the desire to return to aspects of pre-industrial ways of life where village or extended-family based societies predominated. It romanticizes the pre-industrial way of life for its sense of community, for its frugality, and for the fixed place in the social order it attributed to most people. It is a model still found in much of the 'underdeveloped' world, as well as within conscious communities (including some monastic orders) within modern, industrial societies.

Advocates of this model call for the disbandment of existing institutions and a withdrawal of legitimacy from current social arrangements. The 'fourth worlders', as the radical decentralists are sometimes known, believe that the State will simply disappear if people stop attributing to it any power or legitimacy. The New Zealand Green Party's unofficial policy booklet for the 1990 election, *Steps to a Green Land*, represented this type of approach. It argued that community regeneration is the key to economic revival and that the flight of profit-oriented capital from New Zealand, which would inevitably occur once the green economic reforms were implemented, should be welcomed. New Zealand would cancel its economic debts (immediately reducing the nation to the status of a Third World country) and a nebulous promise of community economic activity, barter, and 'an extensive cashless economy' would be the basis of meeting the needs of New Zealanders. A miraculous 'community' would arise, informed and willing to participate in self-governance. Thus the answer to the problem of industrial society '. . . is not to do away with technology, but to do away with the dictates of the dollar and return power back to the communities' (Green Party, 1990: 21).

Another local publication, by Values Party member Bernard Merwood (1989), places decentralization at the top of his list of the core principles of the green movement. Decentralization encourages self-reliance and prevents the capture of power. Furthermore 'only in small units is man safe from himself' (Merwood, 1989: 7). This vision includes the assertion that all land would eventually be transferred to community control and that most governmental decisions would be made at the level of the neighbourhood, town, or borough. Communities would be coordinated by regional governments and communities would receive taxes, administer public services, and be responsible for environmental protection. Mobility, trade, and other activities between communities would be discouraged because of the inevitable waste this involves, whether of energy, because of the transportation involved, or the packaging needed to carry goods over long distances.

These oversimplified utopian visions detract from the real problems which green politics seeks to address, *including* over-centralization and the sense of powerlessness which many people feel in the modern State. There is no model, however, of how the radical changes envisaged by these utopian schemes will occur. From where, for example, is a self-governing and politically active community suddenly going to arise? The search for 'community' has been constant in an inhumanly scaled and highly mobile society, but it cannot be willed into existence by wishful thinking. More importantly, perhaps, what guarantee is there that decision-making at the level of the community is going to be any more just or democratic than decision-making at higher levels? In fact there may be even fewer safeguards for minorities in local and/or direct forms of democracy than in existing political institutions. Particularly in situations where permanent minorities exist, such as the Māori in New Zealand, direct democracy affords few safeguards to ensure that the minority will not always have their interests and views overridden by the majority. Homosexual law reform may never have occurred in New Zealand had such decisions been left up to 'the community', several hundred thousand of whom signed a petition against the decriminalization of homosexual activities. Yet greens speak unequivocally of the belief that 'the decisions of the grass-roots level must, in principle, be given priority' (Spretnak and Capra in Brackley, 1990: 190).

The assumption that grass-roots democracy is egalitarian, and empowers ordinary people, is also questionable. In fact, there may be even fewer checks on a local tyrant or élite than on a larger, more visible, organization. Those greens who aspire to return to village-based systems or who romanticize about the lack of bureaucracy in the political structures of indigenous people

> . . . seldom point out that democracy was conspicuously absent from those bucolic villages they hold up for emulation—villages ruled by the cruellest patriarchy, religious mind-control, feudal ignorance, and force. (Toffler, 1990: 381)

Furthermore, empirical research by political scientists has shown that political élites (political leadership) 'generally take a more liberal line on social and moral issues than the general public' (Hague and Harrop, 1990: 85). Political élites also have more intensely held beliefs, which are more coherent and stable and supported by a wider range of information, than the beliefs of the non-élite. 'In a sense, it is only the members of élites who have *systems* of political belief at all' (Hague and Harrop, 1990: 86). Populist advocates of increased participation like 1992 American presidential contender, H. Ross Perot, claim that people want to be more involved, not because of an inherent interest in politics or the issues but 'because, finally they pay the bills' (*Guardian Weekly*, 7 June 1992: 10). Most greens assume that people want to be involved for a different reason: that they want to have some control over the decisions affecting their own lives. This applies to schools, workplaces, and communities.

But just how keen are people to participate to a greater degree in making decisions about matters beyond their own lives? Models of so-called 'grass-roots democracy' depend upon the willingness of people to participate in the decisions affecting their lives. But as Selbourne suggests, experience of the real political world declares

> that a 'self-confident community', both able and willing in significant numbers— let alone large and active numbers—to exercise such democratic rights and freedoms, cannot be spirited into being by populist incantation. (Selbourne, 1984: 226)

Participatory democratic models have even less saliency at a time of mass unemployment and economic decline such as that which most modern industrial democracies could now be said to be experiencing:

> . . . economic regression, social decline, and above all educational failure are not a politically promising basis for 'new' schemes of 'maximising' democracy or 'devolving' power to 'the people'. (Selbourne, 1984: 208)

Yet calls for an 'active citizenry' (Frankel, 1992) are still promoted as the essential ingredient for the achievement of various green alternatives to modern society. Ekins (1992: i), for example, argues that: 'it is only through the success of democratic popular mobilization that a new world order, based on peace, human dignity and ecological sustainability, can be created'.

Nonetheless, while the new movements seek to 'politicize civil society in ways that are not constrained by representative-bureaucratic political institutions and thereby to reconstitute a civil society independent from increasing control and intervention' (Offe, 1987: 64), contemporary experience with residents' groups and community councils at a local level in New Zealand politics confirms the difficulty of getting people involved in the components of civil society such as local democracy. Where groups do exist, they are frequently not representative of the local community for which they claim to speak and, often, little allowance is made for the views and concerns of different groups. For example, ethnic groups which may not be used to traditional Pākehā meeting structures and processes are often disadvantaged. To create systems of localized decision-making which do not fall prey to an activist minority or to a coterie of local power-brokers is very difficult.

Furthermore, 'local' government has often been the preserve of local business, as empirical studies from political scientists have demonstrated:

> The people who use their time and money to participate in local affairs are the ones who—in vast disproportion to their representation in the population—have the most to gain or lose in land-use decisions . . . (Logan and Molotch, 1987: 620)

The ideal of decentralization is based on a positive view of human nature which accepts that people will behave in the best interests of the entire community when they are put in a position of responsibility. There are plenty of examples of where this is true. There are also far too many examples of where the opposite is true. Different green strategies will be informed by different views about human nature:

> A human race which has genuinely 'internalised' the convictions and disciplines necessary to live within a finite environment might in fact prove to be worshipful, fearful, self-limiting, imaginatively timid, an acceptor of authority, and a searcher after safety. I hope not, but I cannot be sure. (Coombs, 1990: 59)

I do not share the decentralist's faith in the outcomes of local decision-making. Whether in the case of the federal intervention to stop segregation in the south of the United States, or the recent overruling by the central government's Health Minister of a small Italian town's decision to test all new school entrants for AIDS (*Evening Post*, 28 January 1992), it is often centralized bodies which have stepped in to ensure the rights of minorities. Similarly, it is the European Community which has brought the greatest pressure to bear on Britain for its continued poor environmental record, particularly regarding emissions from power stations which have caused acid rain over Scandinavia. Analogously, New Zealand's development as a colonial nation was impeded by the parochial perspective engendered by separate regional governments. The major achieve-ments in New Zealand history, whether in the creation of a welfare state or the pursuit of a nuclear-free policy (not to mention socially progressive legislation for women, Māori, and minorities), were largely the result of enlightened decision-making by central government (often in the face of opposition from large parts of the community, as in the case of homosexual law reform).

Ecotopia embraces rather ambivalent attitudes towards democracy. On the one hand there would be decentralized decision-making, but on the other it is clear that this would take place within very clear boundaries. What of the local community which decides to fell its forest or invite a irradiating food factory into its area to provide jobs for the young unemployed people there? These are serious issues, with applicability to many towns and regions in New Zealand, such as those where the felling of native forests plays an important role in the local economy, like Tuatapere in Southland. Yet the model of decentralization is based on the assumption that local decision-making will be compatible with a green agenda. This ignores some inevitable conflicts between different objectives (for example, job creation versus forest preservation) as well as the unequal distribution of power within local communities.

In short, any vision of a decentralized green society based on the mobilization of an active citizenry is bound to fail because there is little evidence that such mobilization is likely to occur. Greens need to develop more sophisticated change strategies, and to re-evaluate their mistrust of political power.

Green Politics and the Open Society

Ecotopia raises too many imponderables, it poses a threat to human rights, and it fundamentally misunderstands the way in which the environmental crisis—which provides its *raison d'être*—will be resolved. Ecotopia resorts to a restriction of human rights in order to save the world. The green vision of local self-sufficiency is a threat to human rights, if only because of the very scarcity it will encourage:

> The great, necessarily arbitrary, power placed in the hands of political leaders and their bureaucrats in a resource-frugal society constitutes a serious, dangerous loss of freedom and an objectionable source of danger to individuality. (McCloskey, 1983:131)

If we are to learn anything from the history of modern society in the twentieth century (from Nazism, Stalinism, or Pol Pot, for example), then the maintenance of an open society must be a primary goal of green politics, and reforms must take place within the limits which this places on the extent of political action. Given the possibility that continued environmental degradation might be used to validate an autocratic political regime, it is essential that the defence of human rights becomes a core object of green politics. This is essential not just from a human rights viewpoint, but also because it is the open society which is most likely to develop the necessary responses to the environmental crisis:

> The only realistic, feasible avenue to ecological political reform is through the political institutions of an open society that respects human rights. Those who are fortunate enough to live in liberal democratic states can hope to have a much greater measure of success than can those who live in closed societies, even those that sail under the flag of ecological concern, working towards solutions of ecologically based social problems. (McCloskey, 1983: 159)

Saving the world will best be achieved by increased choices in an open society. Already it is clear from a comparison of the environmental standards, policies, and initiatives in different countries, that it is open societies, such as the Scandinavian democracies, which are taking the most concerted action to alleviate the environmental crisis and to pursue sustainable development.

By contrast, the green ecotopia does not represent an 'open society'. The self-sufficient and de-urbanized society is a closed society. It also conflicts with the underlying trend in modern society towards greater diversification and globalization. Postindustrial or postmodern society will be increasingly eclectic in terms of its politics, its lifestyles, and its values. That is a product of the breakdown of the 'mass' tendencies of the industrial era—for example, 'mass politics' (reflected in 'mass parties'), 'mass production', and 'mass religion'. Postmodern society will be characterized by increasing diversity, which will depend upon mutual tolerance and respect. Diversity will be a safeguard against totalitarianism, because a '. . . low diversity society is relatively easy to run from the top' (Toffler, 1983: 96). But, most importantly, a diverse pluralistic system will be more able to respond constructively and innovatively to new issues such as those raised by the environmental crisis. Systems theory notes that a diverse society or institution is more likely to be able to adapt to new challenges, because of its ability to respond to a variety of complex situations, than a closed system with a limited range of inflexible responses (Capra, 1982).

Unresolved environmental problems have the potential to undermine the basis of liberal democracy, as even Francis Fukuyama (1992: 46), in his argument about the inevitable forward march of liberal democracy, alludes to. Escalating environmental crises or disasters could serve to legitimize totalitarian regimes which claim unchecked power in order to 'save the world':

> . . . if ecological scarcity does impact as a sudden emergency . . . then there is every reason to fear an authoritarian-hierarchical social response: perhaps even a neo-fascist alliance of big capitalism and populist reaction, rather than the broadly egalitarian common acceptance of common restrictions that marked World War Two. (Ryle, 1988: 63)

There is a close relationship between predictions of impending ecological catastrophe and calls for an 'ecological dictatorship' (Rüdig, 1985: 21). Several greens around the globe—for example, Pentti Linkola in Finland—have already declared their willingness to adopt the mantle of ecological dictator in their respective countries. Another important factor is the potential for environmental issues to become major sources of international conflict in the future. With the possibility of competition for scarce resources (such as clean air and water) increasing as a result of climate change, it is possible that conflicts over resources might become the major cause of international tension. Worst-case scenarios suggest the possibility of mass-migration from Africa to the European continent as a result of environmental disaster in Africa, while issues of access to fresh water supplies in areas of conflict such as the Middle East could produce an even greater tension between hostile nations.

The inevitable conclusion—for anyone concerned with human rights—from a perusal of modern history, is that reforms towards a sustainable society must be conducted within a democratic framework which respects individual rights. The question that we again return to, then, is whether or not it is possible to reform our political institutions so that we can achieve new patterns of development without eroding basic democratic rights. If we are to avoid future totalitarianism, whether of a green or any other fundamentalist hue, combining macro-level reforms with maintained (and enhanced) individual micro-level freedoms is our best hope of ensuring both necessary policy reforms and social justice.

Green Tactics and Strategies

Given the enormous magnitude of what green politics aspires to achieve, greens need to give much greater thought to their goals, and how such goals might be realized. Green politics is a minority politics, rarely attracting the support of more than 10 per cent of voters. Yet greens often lack a clear idea of how to turn even their small percentage of the vote into real political outcomes. Commoner (1990: 168) writes that the greatest failure of the 'new movements' in the United States 'has been their inability to translate the millions of votes that their combined adherents represent into significant electoral power and thereby elect people to office who will protect their gains and expand them'. Each of the separate issues of the new movements is seen as a modifying force, but not as the motive force of national policy:

> Yet, taken together, and added to the much older labour movement, the issues that the movements represent comprise not only the major aspects of public policy, but its most profound expressions: human rights, the quality of life, health, jobs, peace, survival. (Commoner, 1990: 169)

The realistic alternative approach to the ecotopian withdrawal from modern society is for greens to work through existing institutions to achieve clearly defined short-, medium-, and long-term goals. Such traditional political responses to the crises of modernity may seem paradoxical to a movement born out of a rejection of the bureaucratic and impersonal institutions resulting from the process of modernization. But even if the political sphere were reappropriated '. . . from the institutions that have come to monopolise it and given back to the societal forces and their institutionally unconstrained action' (Offe, 1987: 75), a sustainable society will not be delivered by a strengthened civil society alone. Nor can we rely on a 'new paradigm', widespread 'new mindedness' (Ornstein and Ehrlich, 1989), or a cultural transformation to deliver the green world. Instead green politics must dirty itself in the messy world of the institutions it was originally formed to reject, on the basis that these institutions provide the most opportune vehicle at this point in history to advance the green agenda beyond the green ghettos of civil society.

Alliances and coalitions with other groups will be necessary to achieve certain goals, such as the Green–ALP (Australian Labor Party) Accord which enabled Labor to govern in Tasmania on the basis of support from the Independent Greens, in exchange for implementing key green policies. Greater cooperation is needed between green politics and the wide range of movements with similar goals.

There needs to be an openness to a variety of strategies for achieving specified goals. Green objectives can be met by means other than the direct control or ownership of large parts of the economy which is the traditional socialist approach, epitomized by current Alliance policies in New Zealand. This may mean that traditional activities have to be curtailed to free up resources for environmental objectives, or that public-owned assets have to be privatized in order that sufficient capital can be found to advance sustainability objectives. These options have to be considered if they will achieve the desired green goal.

The spirit of New Zealand's Resource Management Act 1991 provides a useful framework for green decision-making because it focuses on the prevention or alleviation of the adverse environmental effects of an activity, rather than on control and ownership. Green politics should be quite clear about its goals, such as saving water, and then consider all possible means for achieving that goal. In this case it may be water metering (and charging) for each home. Any socially regressive effects of the implementation of water metering can be dealt with through measures such as rates rebates. But by the very act of metering a previously 'free' resource, people's perceptions of just how valuable a resource water is, will change.

Legislation such as the Resource Management Act encourages the establishment of clear environmental standards and the subsequent monitoring and enforcement of environmental performance measures. But the resources needed to adequately monitor and enforce environmental standards are generally insufficient at all levels of government. This situation will be rectified when sustainability is afforded its rightful place at the centre of all decision-making. In the meantime, government can achieve progress by means which do not require substantial funding, by, for example, setting an example through its own activities.

In New Zealand there is also a need for a local equivalent of the American Environmental Protection Authority (EPA) with funding and powers commensurate with the new importance attached to the environment. Acting as an advocate for the environment is the explicit role of no existing governmental body in New Zealand, and yet a 1989 survey of New Zealander's values showed 'that the overwhelming majority of New Zealanders are convinced of the urgent necessity for environmental protection. . . . There is, therefore, the strongest positive climate of opinion for environmental protection. Policy of the most thorough-going nature could be assured of the most positive response from the New Zealand public' (Gold and Webster, 1990: 44).

To be successful, these policy initiatives require that the greens seek to influence public policy at all levels, which inevitably involves the exercise of political power. Yet there is a squeamishness about political power among many greens, which is, as former International Secretary of the Greens, Sara Parkin (1989: 25), states, '. . . odd for a political movement which proposes the most radical redistribution of wealth ever contemplated—sharing it not merely between classes but between continents and with future generations'. The exercise of political power is inevitable, the challenge is for it to be used to achieve green ends. What is true *within* green parties is also true for the broader political environment: 'If the movement continues to deliberately *not* select who shall exercise power, it does not thereby abolish power. All it does is to abdicate the right to demand that those who exercise power and influence will be responsible for it' (Landry *et al.*, 1985: 10).

The experience of many greens in new social movements has been one of dealing with unresponsive, conventional politicians and centralized parties. There is, therefore, a great deal of understandable cynicism within green politics about conventional forms of political organization, as well as with traditional politicians. This cynical attitude towards political power is fuelled by writers like the Hungarian, George Konrad, who, in his book *Anti-Politics*, argues that we must be wary of politicians 'because in all politicians worth their salt there is present, albeit in a more sober form, some of the dynamite that came out in Hitler with such savage brutality; if there were not, they would not have chosen the politician's trade' (Konrad, 1984: 95). The late Petra Kelly used to quote Konrad and was influenced by him in her writings about political power which looked forward to 'abolishing power as we know it' (Kelly, 1987: 69). Bahro articulated similar sentiments, arguing that because of the power component running through the whole of history towards exterminism, 'it may be that we can only survive in conditions free of power' (Bahro, 1986: 148).

But the attainment of political power and influence is essential to achieving green goals. Electing competent people to public bodies and using the skills and resources of those institutions to advance green objectives is vital. While this is possible at a local government level now in New Zealand, the likely introduction of electoral reform in 1996 should see greens elected to Parliament for the first time.

Towards a Postmodern Society

The collective assumptions upon which most modern, industrialized nations have relied are now questioned by growing numbers of people. Providing a new basis for collective social goals will be a difficult task, however. For example, Lasch's (1991) solution to our problems is for a return to the puritan values of the frugal lower middle class who dominated colonial societies. Postmodernism, by contrast, represents 'the final collapse . . . of the political values of the small country town and puritanism' (Turner, 1989: 203). The values of thrift and delayed gratification give way to a new sensibility which emphasizes immediacy, spontaneity, and sensation (Turner, 1989: 203). The postmodernist tolerance of ambiguity and diversity stands in marked contrast to the totalizing schemes envisaged by those like Frankel (1992), who shares with fundamentalists of all hues, a yearning for a return to a 'clear set of values' by which society might be governed.

But the very foundation of postmodernity, according to Heller and Feher (1988: 1), consists of viewing the world as a plurality of heterogeneous spaces and temporalities. A postmodern culture removes the vestiges of a general normative legitimacy for the polity (Turner, 1989: 199). Politically, this implies the end of universalistic schemes like Marxism, while 'the break-down of the grand narrative is a direct invitation to cohabitation among various . . . small narratives' (Heller and Feher, 1988: 5). Postmodernism refuses absolutes, proclaiming a 'war on totality' (Reimer, 1989: 112). Postmodernism recognizes the ways in which ideologies have been used for untenable ends; an awareness that the '. . . trouble with ideological stories that decide how everybody should live is that sooner or later the person who tries to turn the prescription into reality finds it necessary to use force' (Anderson, 1990: 248).

Deep ecologist, Arne Naess, is described as a 'radical pluralist' capable of reacting to other philosophers 'without accepting a final "truth" in any of their systems' (in Fox, 1990: 87). This book is arguing for a 'radical pluralism' within which green politics, if it can clarify its goals, will play a leading role. The idea of 'a socialist society', or 'a green society', is redundant in a postmodern society. What is possible, however, is 'a greener society', where the green agenda is a pivotal component of the ongoing political process which incorporates a variety of inputs from the varied groups and interests in society. 'The incessant growth of the number, volume and dimensions of social issues is an inevitable feature of modernity . . . stemming from the innovations, technological and social changes of modernity itself' (Heller and Feher, 1988: 115). Responses to the environmental crisis, for example, will likely be marked by the creation of more diverse institutions, and more small-scale initiatives (Luard, 1979: 158–9).

The consistent pattern in postmodern society is towards a greater pluralism. While Scandinavian television carrying advertisements in English language can be interpreted as the reflection of a growing cultural homogeneity, in an English speaking country like New Zealand, there are now unparalleled efforts to promote the Māori language. So while it is easy to argue that, at the economic level, the range of options is actually being reduced as the whole world succumbs to the rigour of monetarist economics and the unquestioned assumptions about the desirability of growth, this conclusion makes the same mistakes as the highly visibilist assumptions of the culture generally. That is, there are many developments happening which, while not as conspicuous as major developments, point to a general growth in diversity. For example, having reached a reasonable standard of living in modern, industrial societies on the basis of 'mass production', people begin to seek out non-standardized goods and services. Handmade furniture crafted from recycled demolition materials is now, for example, a growth industry in New Zealand. Even within the investment sphere, there are ethical investment accounts (for example, the Prometheus Foundation in Napier), catering for investors concerned that their money is invested in socially and environmentally desirable ways. The 'informal economy' and arrangements such as 'green dollars' (where people pay for goods and services through reciprocating time or skills) contribute substantially to many national economies and cater for the growing diversity of needs and behaviours.

This postmodern analysis is as relevant to New Zealand as anywhere. The dairy industry which used to produce a handful of varieties now produces a vast range of cheeses to cater for every palate (and pocket!). Centralized, State-run hospitals face crisis at the same time as increasing numbers of New Zealanders seek 'complementary' health care services from naturopaths, acupuncturists, and traditional Māori healers. Mass religions have given way to a wide variety of spiritual paths, from fundamentalist Christianity to transcendental meditation and 'New Age' practices. Politically, there is an unprecedented degree of discontent with the established two-party system, as reflected in the 85 per cent majority for electoral reform in the September 1992 referendum on the issue. A succession of 'minor' parties since the late 1970s—Social Credit, the New Zealand Party, and now the Alliance—have drawn substantial support for populist programmes including promises of greater accountability and an adherence to party manifestos. Increasing political pluralization lends legitimacy to liberal democratic political systems, like New Zealand's, for they are the only political processes which have proved themselves capable of recognizing a variety of norms and values.

Postmodern culture will be a global culture. This trend has many positive advantages, including greater prospects for peace and the ability of the international community to check excesses undertaken within the boundaries of particular states (such as the massacre of students in Tiananmen Square or genocide in the former Yugoslavia). Alongside the trend towards a global culture there is an unparalleled range of activities based around specific communities and the needs of minorities, which is reflected through institutions such as the Waitangi Tribunal and Te Kohanga Reo in New Zealand, and a host of movements for 'self-determination' around the globe.

The coexistence of a wide range of seemingly contradictory and paradoxical movements can be explained by the fact that: 'Modernization, counter-modernization and demodernization must, therefore, be seen as *concurrent processes*' (Berger *et al.*, 1973: 189). The concurrent growth of fundamentalist religion, the green movement, and New Right economics can all be seen as parts of the same process: of a modern society in transition to a postmodern society with an increasing range of behaviours and values. The only constant in future may well be change. It is vital, therefore, that our political institutions are sufficiently flexible to adapt and respond to the ongoing challenges this will create.

Within this eclectic, postmodern political environment, green politics must recognize that it is only one player competing for the values and allegiances of the public. Any aspirations green politics has to impose its world-view across the whole of society are therefore inappropriate for a postmodern future. Greens extend the parameters of political debate by their very presence in the contemporary political environment, but the full potential of green politics' role as a policy-making agency has not been fulfilled. Yet New Zealand manifestly needs green perspectives on policy options. From a green perspective, for example, the goods and services tax (GST) could be deemed to be desirable as it taxes consumption, which greens want to reduce. A green view of the health system, rather than focusing on the provision of large, centralized, techno-logically provisioned hospitals, might emphasize the extent to which growing numbers of people seek 'complementary' (or alternative) health care. Above all, the current narrow economic debate in New Zealand desperately needs a green input, so that issues such as sustainable development are not sidelined but are made central to economic policy.

A green perspective on the economy need not mean a knee-jerk reaction against the market economy. Even if we were to concede that the dictatorship of the market was the root of all contemporary problems, modern attempts to create non-market economies have been far from successful. Green strategies provide little hope that things might change. The movement towards a cooperative economy, the withdrawal of psychic energies from established institutions, and conscious consumerism do not provide a strategy for a fundamental transformation of a market-led economy. In the short term, at least, financial and legislative mechanisms will provide a sound basis for promoting sustainability. The accomplishment of such measures at the earliest opportunity provides a clear and achievable focus for green political energies.

A Concluding Statement

> Everyone with a little imagination can think up alternative worlds. He who
> would be politically relevant must continually ask himself which of these worlds
> is possible. (Berger *et al.*, 1973: 234)

In this book, I have attempted to address issues which are too rarely discussed in the green dialogue. These include the limits to political action in liberal democracies and the extent to which green politics lacks effective strategies or a clear idea of the ends to which its energies are directed. I have also attempted to describe some practical solutions which green politics, within the constraints of liberal democracy, might realistically hope to achieve.

These suggestions amount to far less than a utopian scheme, and would, were I reading this argument for the first time, arouse none of the intellectual excitement which I experienced when I first read Bahro's call to de-industrialize society, for example. But the history of utopian experiments demands that greens be focused on what they can realistically hope to achieve if they are to be politically relevant and to learn from the experience of radical failures too numerous to mention.

While environmental values will be incorporated into decision-making more and more, other anti-modernist components of the green agenda will not be universally embraced. Current environmental problems are life-threatening and are a primary motivation for the current green politics, but green politics is as much a social and spiritual movement as it is an environmental one. Green politics arises in opposition to modernism in all its facets. The basis of the green philosophy is the desire for an holistic order which recognizes the interconnectedness of all phenomena. However, an overwhelming concern for the protection of the natural environment remains pivotal for green politics. This provides a very clear *raison d'être* for green politics which invites the development of a uniquely green set of policies and well thought-out strategies for their attainment.

The desire for a qualitatively different society based on the proponents' particular ideological predilections has been a recurring theme throughout the period of modernity. But such models can no longer be imposed on the heterogeneous publics of modern societies, such as New Zealand in the 1990s. The achievement of a sustainable society is, however, a realizable goal which can be accomplished within the boundaries of liberal democracy, thereby providing green politics with a very clear focus for its energies which is a constructive alternative to the fanciful pursuit of utopian fantasies. It is easy to provide every home with access to a waste recycling scheme. It is not easy to provide every home with access to a spiritual path.

Nevertheless, green politics also fulfils an educative role, promoting change in individuals' behaviour because of the impact peoples' daily choices have on the kind of society we have. Change at this level will be vital, but it will not be sufficient on its own to achieve the sustainable society at the heart of green aspirations. All levels of government must be involved in creating policy-making frameworks which take the environment and the rights of future generations into account; in short, the primary responsibility of government now is the attainment of a sustainable society.

If green politics wants to play an effective role in advancing this agenda in an eclectic political future, it will have to review its internal *modus operandi*. The late Petra Kelly stated that the greens have 'not yet solved the problem of how to deal with committed, energetic and credible personalities' (Kelly, in Brackley, 1990: 199). Green parties will have to develop efficient structures which, while flexible enough to accomodate the individualists who inhabit green parties (Kitschelt, 1988: 127), also allow for maximum effectiveness. As sectors of the business community increase the chorus of anti-green sentiments, the need for greens to be effective and to advance credible policies will grow.

If green politics is to be successful in future it must learn from its ideological forebears. In New Zealand, where the Values Party was first formed more than twenty years ago, there is an ideal opportunity to learn from history and to develop a credible, focused, and well-organized green presence in the political arena. The alternative will be unrealistic expectations resulting in disappointment and frustration. Many greens see membership of the Alliance as a realistic response to the greens' predicament under the current electoral system. I do not share their views, but hope that the earliest implementation of electoral reform— under which a separate green party would be likely to contest the elections—will resolve the issue of the greens' membership of the Alliance.

The challenge for the greens is to remain faithful to the anti-modernism which is the central thread of their ideology, without abandoning the positive aspects of modernism. Modernity has been liberating for many. For others, it is that from which liberation is sought (Berger *et al.*, 1973: 195). The greens must seek to recapture useful parts of the past without discarding the progress which has been made for many people in modern industrial societies in terms of increased standards of living and enhanced personal freedoms. This requires a sophisticated analysis of exactly what kind of society greens desire, and how to achieve such a society. Are the positive aspects of modernity, such as individual freedoms, compatible with the restrictions on some personal freedoms (such as mobility) implicit in the green agenda? To what extent are we as a society prepared to sacrifice some individual freedoms in pursuit of the greater (collective) environmental good? These are significant questions which not just the green movement but all of modern society needs to address.

At the core of this debate is the question of whether green politics will contribute to an eclectic, postmodern future, or whether it will join the chorus of divergent groups seeking a return to their particular set of values as the basis for guiding all society. Modernity has coupled the liberation of the individual with the creation of anonymous structures, like the State. Some antimodernist movements seek an end to anonymity and abstractness and a return to greater collective solidarity *'even if the price for this should be less autonomy for the individual'* (Berger *et al.*, 1973: 198—my emphasis). Some sectors of the green movement adhere to this view but, like all attempts 'to force the messy actuality of human society into the geometrical mould of an abstract idea' (Scruton, 1985: 74), they are bound to fail. The solutions to the crisis of modernity, including the environmental crisis, will best be achieved through an open society based on a wide range of views and values. In this task, green politics can play a potentially vital role. The effectiveness of green politics in swinging the balance from development towards sustainability will be one of the determinants of whether or not modern society successfully adapts to a postmodern era.

As much as anything, this book has been a plea for human rights, because future battles over resources could well lead to attempts to reduce human rights, (as well as to wars and other conflicts) but also because an environmentally pristine and sustainable society without human rights would be an intolerable place to live. If human rights are to be sustained, however, our society must begin to move towards a democratic consensus on the need for a radically new approach to decision-making which puts environmental sustainability at the centre of all decisions. Our social, economic, and political institutions must begin now to mobilize their collective resources towards the goal of achieving a sustainable society. It is therefore the prime responsibility of green politics to advance these new goals at all levels of government, as well as within the institutions of civil society. New Zealand is uniquely placed to promote sustainable development because of its relatively unspoiled environment, its isolation, size, and 'clean, green' image. We could become a model to the world, if only our political institutions would have the courage to develop an independent line in this regard, as they have done in the field of foreign policy.

The greatest challenge therefore remains whether or not our political institutions are able to adapt to an age of sustainable development. If the prediction that the transition to sustainable development will be difficult is true, then the protection of human rights and political freedoms will be paramount. Green politics may have to be prepared to reduce the focus of its holistic attentions, rooted in a wide range of humanist and counter-culture movements from the last one hundred years, to the defence of these fundamental rights, as well as to immediate achievement of a consensus about the creation of a sustainable society upon which the future of the planet depends.

Bibliography

Andersen, J.G., 1990. 'Environmentalism, New Politics, and Industrialism: Some Theoretical Perspectives', *Scandinavian Political Studies* 13 (2):101–18.

Anderson, Terry L., and Leal, Donald R., 1991.*Free Market Environmentalism*, Boulder, Westview Press.

Anderson, Walter Truett, 1990. *Reality Isn't What It Used To Be*, San Francisco, Harper and Row.

Bahro, Rudolf, 1984. *From Red to Green*, London, Verso.

Bahro, Rudolf, 1986. *Building the Green Movement*, London, GMP Publishers.

Baker, K.L., Dalton, R.J., and Hildebrandt, K., 1981. *Germany Transformed*, Cambridge (Mass.), Harvard University Press.

Bauman, Z., 1989. *Modernity and the Holocaust*, New York, Cornell University Press.

Bell, Daniel, 1982. 'The Return of the Sacred', in Almond, G. A., *et al.* (eds), *Progress and Its Discontents*, Berkeley, University of California.

Berger, Peter, *et al.*, 1973. *The Homeless Mind*, New York, Vintage.

Berman, M., 1980. *The Politics of Authenticity*, New York, Atheneum.

Berman, M., 1981. *The Re-enchantment of the World*, Ithaca, Cornell University Press.

Blackwell, T., and Seabrook, J., 1988. *The Politics of Hope*, London, Faber.

Boulding, K.E., 1978. *Ecodynamics: A New Theory of Social Evolution*, Beverley Hills, SAGE.

Brackley, Peter (ed.), 1990. *World Guide to Environmental Issues and Organizations*, London, Longman.

Broad, Harry, 1991. 'Tunes from an Economic Instrument', *Terra Nova* 4 (April): 19–22.

Brooks, Paul, 1989. *The House of Life*, Boston, Houghton Mifflin.

Brundtland, G–H., *et al.*, 1987. *Our Common Future*, New York, UNEP.

Brundtland, G–H., 1990. 'Epilogue', in Scientific American, *Managing Planet Earth*, New York, W.H. Freeman.

Brunt, A., 1973. 'In Search of Values', in Edwards, B. (ed.), *Right Out*, Wellington, Reed.

Buber, Martin, 1958. *Paths in Utopia*, Boston, Beacon.

Callenbach, Ernest, 1975. *Ecotopia*, Berkeley, Banyan Tree Books.

Campbell, Gordon, 1992. 'Last Chance for Planet Earth', *Listener and TV Times*, 24 February: 14–17.

Campbell-Bradley, Ian, 1987. *Enlightened Entrepreneurs*, London, Weidenfield and Nicholson.

Capra, Fritjof, 1982. *The Turning Point*, London, Fontana.

Caute, David, 1988. *Sixty-Eight*, London, Hamish Hamilton.

Chapman, Robert, 1981. 'From Labour to National', in Oliver, W. H. (ed.), *The Oxford History of New Zealand*, 1st edn, Wellington, Oxford University Press.

Cleveland, Les, 1972. *The Anatomy of Influence*, Wellington, Hicks Smith.

Cleveland, Les, 1979. *The Politics of Utopia*, Wellington, Methuen.

Commoner, Barry, 1990. *Making Peace With the Planet*, London, Victor Gollancz.

The Contractor, November 1992: 9–11.

Coombs, H.C., 1990. *The Return of Scarcity*, Sydney, Cambridge University Press.

Cousins, Norman, 1981. *Human Options*, New York, Berkley.

Critic, 6 May 1975.

Croome, Rodney, 1990. 'At the Crossroads: Gay and Green Politics', in Pybus, C., and Flanagan, R., (eds), *The Rest of the World is Watching*, Sydney, Pan.

Dahrendorf, Ralf, 1988. *The Modern Social Conflict*, London, Weidenfield and Nicolson.

Dalton, R.J., Flanagan, S.C., and Beck, P.A. (eds), 1984. *Electoral Change in Advanced Industrial Societies*, Princeton, Princeton University Press.

Daly, Herman, and Cobb, John, 1989. *For the Common Good*, Boston, Beacon Press.

Davis, Peter, 1991. 'A Sense of Belonging', *Greenstone* 2 (November–December): 8–9.

Davis, Peter, and Hodge, Judith, 1990. *The New Zealand Green Guide*, Auckland, Penguin.

Deutsch, K.W., 1985. 'The Systems Theory Approach as a Basis for Comparative Research', *International Social Science Journal* XXXVII (1): 5–18.

Dobson, Andrew, 1990. *Green Political Thought*, London, Unwin Hyman.

The Dominion, 18 March 1992.

The Dominion, 30 April 1992.

The Dominion Sunday Times, 5 April 1992.

Drucker, P.F., 1969. *The Age of Discontinuity*, New York, Harper and Collins.

Dryzek, J.S., 1987. *Rational Ecology*, New York, Basil Blackwell.

Eckersley, Robyn, 1989. 'Green Politics and the New Class: Selfishness or Virtue?', *Political Studies* XXXVII: 205–23.

Eckersley, Robyn, 1990. 'The Ecocentric Perspective', in Pybus and Flanagan (eds), op. cit.

Ehrlich, Anne, and Ehrlich, Paul, 1990. *Extinction*, New York, Ballantine.

Eisler, Riane, 1987. *The Chalice and the Blade*, San Francisco, Harper and Collins.

Ekins, Paul, 1992. *A New World Order*, London, Routledge.

Environment, Science and Technology, 1983, 17 (11): 522–28.

The Evening Post, 16 May 1978.

The Evening Post, 28 January 1992.

The Evening Post, 29 February 1992.

The Evening Post, 12 June 1992.

The Evening Post, 12 December 1992.

Ferguson, M., 1980. *The Aquarian Conspiracy*, London, Paladin.

Fox, Warwick, 1990. *Toward A Transpersonal Ecology*, London, Shambhala.

Friberg, M., and Hettne, B., 1985. 'The Greening of the World—Towards a Non-Deterministic Model of Global Process', in Addo, H., *et al.*, *Development as Social Transformation*, London, Hodder and Stoughton.

Fromm, Erich, 1976. *To Have or To Be?*, London, Abacus.

Fukuyama, Francis, 1992. *The End of History and The Last Man*, London, Penguin.

Gaiter, P.J., 1991. *The Swedish Green Party*, Stockholm, Stockholm University.

Galbraith, J.K., 1977. *The Age of Uncertainty*, London, BBC Books.

Galtung, J., 1988. 'The Green Movement', in Pitt, D. C. (ed.), *The Future of the Environment*, London, Routledge.

Giddens, Anthony, 1990. *The Consequences of Modernity*, Stanford, Stanford University Press.

Giddens, Anthony, 1991. *Modernity and Self-Identity*, Oxford, Polity Press.

Gold, H., and Webster, A., 1990. *New Zealand Values Today*, Palmerston North, Alpha.

Goodwin, B., and Taylor, K., 1982. *The Politics of Utopia*, London, Hutchinson.

Gorz, Andre, 1982. *Farewell to the Working Class*, London, Pluto.

Gorz, Andre, 1989. Interview in *New Statesman and Society*, 12 May 1989: 27–31.

Gould, P.C., 1988. *Early Green Politics*, Sussex, Harvester.

Green Party of New Zealand/Aotearoa, 1990. *Steps to a Green Land*, election manifesto.

Greenstone, 1992. 3 (December).

Guardian Weekly, 7 June 1992.

Gustafson, Barry, 1980. *Labour's Path to Political Independence*, Auckland, Oxford University Press.

Hague, R., and Harrop, M., 1990. *Comparative Government and Politics*, London, Macmillan.

Hayward, M., 1981. *Diary of the Kirk Years*, Wellington, Reed.

Heller, Agnes, and Feher, Ferenc, 1988. *The Postmodern Political Condition*, Oxford, Polity Press.

Howard, Dick, 1989. *Defining the Political*, London, Macmillan.

Hülsberg, W., 1988. *The German Greens*, London, Verso.

Hughes, H. Stuart, 1990. *Sophisticated Rebels*, Cambridge (Mass.), Harvard University Press.

Hutton, Drew, 1987. *What is Green Politics?*, Sydney, Angus and Robertson.

James, Colin, 1992, *New Territory*, Wellington, Bridget Williams Books.

Jennett, Christine, and Stewart, Randal G., (eds), 1989. *Politics of the Future*, Melbourne, Macmillan.

Katsiaficas, George, 1987. *The Imagination of the New Left*, Boston, South End Press.

Kelly, Petra, 1987. 'Towards a Green Europe! and a Green World!', *Miljöpartiet de Gröna/Green Thinking for Global Linking*, Stockholm.

Kitschelt, H., 1988. 'Organisation and Strategy of West German and Belgian Ecology Parties', *Comparative Politics*, January: 127–54.

Kolinsky, Eva, 1989. *The Greens in West Germany*, Oxford, Berg.

Konrad, George, 1984. *Antipolitics*, London, Quartet.

Landry, C., Morley, D., Southwood, R., and Wright, P., 1985. *What a Way to Run a Railroad*, London, Comedia.

Larkin, Andrew, 1988. 'The Ethical Problem of Economic Growth in Environmental Degradation', in Schrader-Frechette, K.S., *Environmental Ethics*, California, Boxwood Press.

Lasch, C., 1991. *The True and Only Heaven*, New York, Morton.

Lawson, Kay, 1976. *The Comparative Study of Political Parties*, New York, St Martin's Press.

Lawson, Kay, 1988. 'When Linkage Fails', in Lawson, K., and Merkl, P., (eds), *When Parties Fail*, Princeton, Princeton University Press.

Leach, R., 1988. *Political Ideologies, An Australian Introduction*, Sydney, Macmillan.

Lee, Keekok, 1989. *Social Philosophy and Ecological Scarcity*, London, Routledge.

Levine, S., 1975. *New Zealand Politics: A Reader*, Melbourne, Cheshire.

Logan, J.R., and Molotch, H.L., 1987. *Urban Fortunes*, Berkeley, University of California Press.

Lohrey, Amanda, 1990. 'The Greens: A New Narrative', in Pybus and Flanagan (eds.), op. cit.

Luard, E., 1979. *Socialism Without the State*, London, Macmillan.

McCloskey, H.J., 1983. *Ecological Ethics and Politics*, New Jersey, Rowman and Littlefield.

Mackenzie, Dorothy, 1991. *Green Design*, London, Laurence King.

Mackintosh, Maureen, and Wainwright, Hilary, 1987. *A Taste of Power: The Politics of Local Economics*, London, Verso.

McLeod, Marion, 1992. 'Looking for Utopia', *Listener and TV Times*, 27 January: 26–27.

MacNeill, Jim, 1990. 'Strategies for Sustainable Economic Development', in *Scientific American*, *Managing Planet Earth*, op. cit.

McVarish, Scott, 1992. *The Greening of New Zealand*, Auckland, Random Century.

Maier, Charles S., 1987. *Changing Boundaries of the Political*, Cambridge, Cambridge University Press.

Manes, Christopher, 1990. *Green Rage*, Boston, Little Brown.

Marien, M., 1986. 'The Transformation as Sandbox Syndrome', in May, R., *et al.*, *Politics and Innocence*, Dallas, Saybrook.

Martineau, Alain, 1986. *Herbert Marcuse's Utopia*, Montreal, Harvest House.

Melucci, Albert, 1989. *Nomads of the Present*, London, Hutchinson.

Merwood, Bernard, 1989. *Green Politics for New Zealand*, Bernard Merwood, Warkworth.

Miller, Alan, 1991. *Gaia Connections*, Rowman and Littlefield.

Mishan, E.J., 1986. *Economic Myths and the Mythology of Economics*, Brighton, Wheatsheaf.

Müller-Rommel, F., 1985. 'The Greens in Western Europe, Similar But Different', *International Political Science Review* 6 (4): 483–99.

Naisbitt, John, 1982. *Megatrends*, London, Macdonald.

Nicholson-Lord, David, 1987. *The Greening of the City*, London, Routledge and Kegan Paul.

Nisbet, Robert, 1980. *History of the Idea of Progress*, London, Heinemann.

Offe, Claus, 1984. *Contradictions of the Welfare State*, London, Hutchinson.

Offe, Claus, 1987. 'Challenging the Boundaries of Institutional Politics: Social Movements Since the 1960s', in Maier, op. cit.

Olofsson, G., 1988. 'After the Working Class Movement? An Essay on What's New and What's Social About the New Social Movements', *Acta Sociologia* 31 (1): 15–34.

Ophuls, W., 1977. *Ecology and the Politics of Scarcity*, San Francisco, Freeman.

Ornstein, R., and Ehrlich, P., 1989. *New World New Mind*, New York, Simon and Schuster.

Paehlke, R., and Torgerson, D., 1990. *Managing Leviathan: Environmental Politics and the Administrative State*, London, Broadview Press.

Pakulski, Jan, 1990. *Social Movements, The Politics of Moral Protest*, Melbourne, Longman Cheshire.

Palmer, Geoffrey, 1990. *Environmental Politics*, Dunedin, John McIndoe.

Papadakis, Elim, 1984. *The Green Movement in West Germany*, Kent, Croom Helm.

Papadakis, Elim, 1989. 'Green Issues and Other Parties: *Themenklau* or New Flexibility?', in Kolinsky, op. cit.

Parkin, S., 1989. *Green Parties: An International Guide*, London, Heretic.

Pepper, David, 1985. 'Determinism, Idealism and the Politics of Environmentalism—A Viewpoint', *International Journal of Environmental Studies* 26: 11–19.

Pepper, David, 1991. *Communes and the Green Vision*, London, Green Print.

Phipps, John-Francis, 1990. *The Politics of Inner-Experience*, London, Green Print.

Pierson, S., 1979. *British Socialists, The Journey From Fantasy to Politics*, Cambridge (Mass.), Harvard University Press.

Polsgrove, Carol, 1991. 'Search For Security', *The Progressive*, May: 38–41.

Porritt, J., and Winner, D., 1988. *The Coming of the Greens*, London, Fontana.

Porter, G., and Brown, J. Welsh, 1991. *Global Environmental Politics*, Boulder, Westview.

Rain, Lynn, 1991. *Community, The Story of Riverside*, Nelson, Riverside Trust.

Rainbow, Stephen, 1991. 'Is History About to Repeat Itself? The Greens and the 1990 New Zealand General Election', in McLeay, E. M. (ed.), *The 1990 General Election: Perspectives on Political Change in New Zealand*, Wellington, Department of Politics, Victoria University of Wellington.

Rainbow, Stephen, 1992. 'Why Did New Zealand and Tasmania Spawn the World's First Green Parties?', *Environmental Politics*, 1 (3): 321–46.

Reimer, Bo, 1989. 'Postmodern Structures of Feeling: Values and Lifestyles in the Postmodern Age', in Gibbins, John (ed.), *Contemporary Political Culture*, London, SAGE.

Riddell, Carol, 1990. *The Findhorn Community*, Scotland, Findhorn Press.

Robertson, James, 1990. *Future Wealth*, London, Cassell.

Roddick, Anita, 1991. *Body and Soul*, London, Ebury Press.

Roth, Roland, 1991. 'Local Green Politics in West German Cities', *International Journal of Urban and Regional Research* 15 (1): 75–89.

Ruckelshaus, W.D., 1990. 'Towards a Sustainable World', in Scientific American, *Managing Planet Earth*, op. cit.

Rüdig, W., 1985. 'The Greens in Europe: Ecology Parties and the European Elections of 1984', *Parliamentary Affairs*, 38 (1): 56–72.

Rüdig, W., and Lowe, P.D., 1986. 'The Withered "Greening" of British Politics: A Study of the Ecology Party', *Political Studies* XXXIV (2): 262–84.

Ryle, Martin, 1988. *Ecology and Socialism*, London, Radius.

Salmon, Guy, 1991. 'There is a place for people in nature', *The Dominion*, 16 May 1991: 10.

Schneiderman, Leo, 1988. *The Psychology of Social Change*, New York, Human Sciences Press.

Schor, Juliet, 1991. *The Overworked American*, New York, Basic Books.

Schrader, Ben, 1991. *A Peaceful Path to Reform*, unpublished History honours paper, Victoria University of Wellington.

Schrecker, Ted, 1990. 'Resisting Environmental Legislation: The Cryptic Pattern of Business-Government Relations', in Paehlke and Torgerson, op. cit.

Scruton, Roger, 1985. *Thinkers of the New Left*, London, Longman.

Seabrook, Jeremy, 1990. *The Myth of the Market*, Devon, Green Books.

Selbourne, D., 1984. *Against Socialist Illusion*, London, Macmillan.

Seyd, Patrick, 1987. *The Rise and Fall of the Labour Left*, London, Macmillan.

Sheehy, Gail, 1990. *The Man Who Changed the World*, New York, Harper Collins.

Short, John, 1989. *The Humane City*, Oxford, Basil Blackwell.

Shrader-Frechette, K.S., 1981, *Environmental Ethics*, (Pacific Grove, Boxwood Press).

Spretnak, Charlene, 1986. *The Spiritual Dimension of Green Politics*, Santa Fe, Bear and Co.

Spretnak, Charlene, and Capra, Fritjof, 1986. *Green Politics*, Santa Fe, Bear and Co.

Starr, Paul, 1992. 'A Challenge for Liberalism', *Dialogue* 97 (3): 37–43.

Steinem, Gloria, 1992. *Revolution From Within*, London, Bloomsbury.

Stockwin, J.A.A., 1982. *Japan: Divided Politics in a Growth Economy*, New York, Norton.

Suzuki, David, 1990. *Inventing the Future*, Sydney, Allen and Unwin.

Time, 30 March 1992: 51.

Toffler, Alvin, 1980. *The Third Wave*, London, Collins.

Toffler, Alvin, 1983. *Previews and Premises*, New York, William Morrow.

Toffler, Alvin, 1990. *Power Shift*, New York, Bantam.

Turner, Bryan S., 1989. 'From Postindustrial Society to Postmodern Politics: The Political Sociology of Daniel Bell', in Gibbins, J. R. (ed.), *Contemporary Political Culture*, London, SAGE.

Values Party, 1972. *Blueprint for New Zealand*, election manifesto.

Values Party, 1975. *Beyond Tomorrow*, election manifesto.

Values Party Archives, WTU 85/11, Box 10, Alexander Turnbull Library, National Library of New Zealand.

Vedung, E., 1989. 'Green Lights for the Swedish Greens', pre-published paper, Uppsala University.

Victoria University of Wellington, 1991. 'World Agency Needed to Protect Environment', *Research Report 1991*: 38–40.

Young, Nigel, 1977. *An Infantile Disorder*, Boulder, Westview Press.

Zweig, Paul, 1980. *The Heresy of Self-Love*, New Jersey, Princeton University Press.

Index